In Principle, In Practice

LEARNING INNOVATIONS SERIES

SERIES EDITORS
John H. Falk
Lynn D. Dierking

BOOKS IN THE SERIES

In Principle, In Practice: Museums as Learning Institutions, edited by John H. Falk, Lynn D. Dierking, and Susan Foutz

ABOUT THE SERIES

Museums, libraries, broadcast and print journalism, the Internet, and community-based organizations all represent distinct and important sources of public education that learners can choose from at will. The Learning Innovations Series publishes books pertaining to this broadly defined area of free-choice learning for use by free-choice learning educators and researchers, in college and university courses or by scholars and professionals in the social sciences and humanities. Each volume in the series focuses on a different institution or subject area, highlighting the many ways in which the free-choice sector facilitates learning in our society.

ABOUT THE ORGANIZATION

Established in 1986 as a not-for-profit learning research and development organization, the Institute for Learning Innovation is dedicated to changing the world of education and learning by understanding, facilitating, advocating and communicating about free-choice learning across the life span. The Institute provides leadership in this area by collaborating with a variety of free-choice learning institutions such as museums, other cultural institutions, public television stations, libraries, community-based organizations such as scouts and the YWCA, scientific societies and humanities councils, as well as schools and universities, striving to better understand, facilitate and improve their learning potential by incorporating free-choice learning principles.

In Principle, In Practice

Museums as Learning Institutions

Edited by John H. Falk,
Lynn D. Dierking,
and Susan Foutz

A Division of
ROWMAN & LITTLEFIELD PUBLISHERS, INC.
Lanham • New York • Toronto • Plymouth, UK

AltaMira Press
A division of Rowman & Littlefield Publishers, Inc.
A wholly owned subsidiary of The Rowman & Littlefield Publishing Group, Inc.
4501 Forbes Boulevard, Suite 200
Lanham, MD 20706
www.altamirapress.com

Estover Road
Plymouth PL6 7PY
United Kingdom

British Library Cataloguing in Publication Information Available

Library of Congress Cataloging-in-Publication Data

In principle, in practice : museums as learning institutions / edited by John H. Falk, Lynn D. Dierking, and Susan Foutz.
p. cm.—(Learning innovations)
ISBN-13: 978-0-7591-0976-6 (cloth : alk. paper)
ISBN-10: 0-7591-0976-1 (cloth : alk. paper)
ISBN-13: 978-0-7591-0977-3 (pbk. : alk. paper)
ISBN-10: 0-7591-0977-X (pbk. : alk. paper)
1. Museums—Educational aspects. I. Falk, John H. (John Howard), 1948–II. Dierking, Lynn D. (Lynn Diane), 1956–III. Foutz, Susan, 1979–

AM7.I43 2007
069.1—dc22 2006102098

Printed in the United States of America

∞™ The paper used in this publication meets the minimum requirements of American National Standard for Information Sciences—Permanence of Paper for Printed Library Materials, ANSI/NISO Z39.48-1992.

Contents

Foreword

David A. Ucko

Most people, most of the time, learn most of what they know outside the classroom.

—George Tressel, former division director,
Materials Development, Research,
and Informal Science Education, NSF

Although people have always learned outside the four walls of a classroom, the purposeful design of experiences for self-directed learning by organizations established with that mission expanded greatly in the late 20th century. That increase is best illustrated by the growth of the Association of Science-Technology Centers (ASTC) from 20 museums at its inception in 1973 to some 340 U.S. members today. These institutions enable 83 million citizens each year to experience science and technology firsthand, mostly as families and school groups.

Not surprisingly for a young field, informal education has largely been guided by practice informed by personal experience and intuition. Evaluation and research are only now beginning to play greater roles. One reason is the growing demand for accountability by public and private funders who are seeking evidence for their investments in informal education. Other factors stem from maturation of the field. The value of formative evaluation in developing exhibits and programs is gaining acceptance, as is the potential to learn from what does and doesn't work based on summative evaluation. In addition, increasing numbers of researchers in academic institutions as well as some museums have been studying how people learn in informal settings and are publishing their results in peer-reviewed journals. All of these factors are helping to professionalize the field.

The Informal Science Education (ISE) program and its predecessors at the National Science Foundation (NSF) have supported the field for more than two decades. They were instrumental in providing early funding for capacity building within ASTC. ISE has invested heavily in exhibitions, as well as in a wide range of educational programs. Its funds have made possible more than half of the approximately 200 exhibitions that have been toured by the ASTC Traveling Exhibition Service. NSF was instrumental in establishing the field of children's science programming on television through support of *3-2-1 Contact*, *Bill Nye*, and *The Magic School Bus*, and adult science programming through funding for *NOVA* and the National Public Radio science unit. ISE investments established the large-format film as an immersive educational medium and made possible such out-of-school programming as "citizen science," in which the public contributes to ongoing scientific research.

While funding development of products designed to increase public interest, engagement, and understanding of science and technology, ISE has elevated standards through program emphasis on accuracy of scientific content, linkages to formal education, reaching underserved audiences, and evaluation. Recently, ISE revised its solicitation to require that projects seek to "raise the bar" by furthering knowledge or practice ("strategic impact"), as well as demonstrating innovation and collaboration. To meet these requirements, proposers—principal investigators, or PIs in NSF lingo—must be aware of and build upon the lessons learned from both prior practice and educational research. ISE has invested in several projects to help PIs advance the field in this way.

The website www.informalscience.org at the University of Pittsburgh Center for Learning in Out of School Environments (UPCLOSE) provides access to a searchable database of educational research articles along with both formative and summative evaluation studies from ISE-funded projects. Conferences such as *Best Practices in Science Exhibition Development* and *Web Designs for Interactive Learning* have helped enlarge and share information. A synthesis study now underway at the National Research Council will assess and report on what is known about the characteristics of effective informal science education environments across a range of outcome measures. In addition, the program this year intends to fund an Informal Science Education Resource Center designed to stimulate further professionalization and a community of practice that bridges the field.

The *In Principle, In Practice* initiative, which includes a national conference that was organized by the Institute for Learning Innovation along with a series of publications, forms an important part of this larger ISE effort. Following a synthesis of research findings published in a supplement to the journal *Science Education*, the conference explored and articulated explicit connections among research, evaluation, and practice. The resulting brief reports and this book now share these conversations with a wider audience. Their intent is to identify critical issues in museum learning, better establish a base of research evidence, improve practice in the field, and guide future research efforts. By achieving these outcomes, this initiative will augment the value of informal or free-choice learning on its own terms, rather than those derived from formal education, which do not closely match our community's unique strengths, especially outside the cognitive domain. In so doing, it should also enhance the credibility of the informal education field among policymakers.

The primary focus for *In Principle, In Practice* has been learning in science museums. However, many outcomes may be more broadly applicable to other types of museums, to modes of informal science education other than museums, and to learning in general. This conference fostered active exchange between researchers and practitioners about the ideas that are discussed in the pages that follow. It is essential to the further development of the entire informal education community that this dialogue continues and expands.

Preface

John H. Falk, Lynn D. Dierking, and Susan Foutz

In May 2005, more than 150 people gathered in a small room off an obscure conference center hallway in Indianapolis. Everyone there was a museum professional, but from diverse parts of the community. There were professionals from art, history, and science museums, and also zoos, botanical gardens, and aquariums. They included educators, curators, and administrators, as well as researchers, evaluators, exhibition designers, and funders, professionals from large institutions and ones from smaller institutions. And since this was a session at an annual meeting of the American Association of Museums, most did not know each other or did not expect to get to know each other over the next hour and 15 minutes. Within ten minutes, though, all of these individuals had self-organized into roughly a dozen discussion groups and were deeply engaged in conversation and debate. Although the topics had been predefined, everyone seemed able to find something of interest among questions that ranged from how to make collections more relevant to determining the long-term impact of museum experiences to improving the role of staff in facilitating visitor learning. As new people joined, they were quickly given an overview of the topics and sent to find a group; they too became enmeshed in conversation and debate. There were small groups everywhere, including a couple of groups that moved their tables and chairs out into the hallway. Although these professionals had not known each other at the beginning of the session and came from varied backgrounds and institutions, they all had something in common. They shared a thirst to know more about the museum community in which they worked and to learn what others were thinking were important topics in the community, and they possessed a sincere desire to use this information to support and enhance the public learning that happens in museums of all types. These are the very reasons that motivated us to create this edited volume, and we suspect they are the same reasons that have motivated you to read this book.

To understand and make the best use of this book, though, you should know that there's a story behind the volume. More than a decade ago, in 1994, with the support of the National Science Foundation, the Institute for Learning Innovation (then Science Learning, Inc.) hosted a national conference in Annapolis, Maryland—*Public Institutions for Personal Learning: Establishing a Long-Term Research Agenda.* The goal of the conference was to discuss the nature of museum learning and to formulate a research agenda to investigate the long-term impact of visits to museums. This effort, in particular the edited volume that resulted from the conference (Falk & Dierking, 1995), served as a catalyst for numerous museum research endeavors, stimulated dozens of masters' theses and doctoral dissertations, and continues to guide research and theory in the field.

The past decade has been a time of significant research and great advances in our understanding of learning in and from museums. Despite this progress, though, the museum community continues to struggle to meaningfully document the impact of its exhibitions, media, community-based programs, websites, and other educational efforts, and to apply those findings to the creation of useful and valid frameworks for exemplary practice. To move this process forward, to consolidate our understandings from a decade's museum-learning research efforts, and to lay out the issues that need to be addressed in the decade ahead, the institute, again with National Science Foundation support, launched a new initiative called *In Principle, In Practice: A Learning Innovation Initiative on Museum Learning.* This initiative had several parts, including a preconference special issue of the journal *Science Education* (July 2004, Volume 88, Supplement 1); a national conference of over 100 museum professionals representing a diverse cross-section of roles and responsibilities, institutions, geography, and, of course, perspective; the national town hall meeting at the American Association of Museums annual meeting described above; production of a set of short discussion papers related to current practical issues and concerns published online as *Insights* (Stein, Dierking, Falk, & Ellenbogen, 2006); and now this book.

Our goal throughout this initiative was to gather together in a single place an accessible compendium of what we currently know about learning in museums, to inquire where that knowledge leads us in terms of practice and community, and finally to ponder what still needs to be learned as the museum community moves ever further into the uncertainties and challenges of a new millennium. This book, then, is actually the product of a very long process. These ideas are the distillation of more than two years of effort to bring a useful volume to the museum community, useful being defined here as one that provokes thought and discussion within the field. Most importantly, the ideas presented here are not merely the products of the authors listed but actually represent the collected ideas of a large sampling of the museum community. As the following quotes from professionals who were interviewed in connection with the *In Principle, In Practice* national conference illustrate, there is a very real need to connect museum research to practice and to articulate the place of museums in society at large:

> The next decade of museum learning research needs to focus on creating a conduit between researchers and practitioners with the primary goal of affecting practice (of course this conference is designed to achieve this). To do so, research must be collaborative and cross disciplinary, and include learning theorists, sociologists, anthropologists, museum evaluators, and exhibition designers, interpreters, developers, and educators from the museums in which the research is being conducted. Learning research in museums must connect findings to the stimuli if museums are to learn how to create more meaningful experiences. (Randi Korn, Randi Korn Associates)

> We certainly have a greater [theoretical] understanding than we used to have 10 to 20 years ago. But that has still not been translated into the different ways in which people design informal science education experiences. Not all practitioners have knowledge of this research. And for those who do, it's not obvious how to translate it into specific applications. . . . [Professionals are] not necessarily learning from evaluations that are done elsewhere or even from their own institution's evaluations done in previous years. So the field needs to do much better in terms of building on resources based on prior experience. It's the only way for a field to move forward. (David Ucko, National Science Foundation)

> We're talking about building learning communities . . . [it is] essential for a healthy democracy and civic well being to have educated citizens. Museums have to see how they fit into the constellation that includes schools, libraries, and homes and how they work with these venues to increase choices for free-choice learning. (Marsha Semmel, Institute for Museum and Library Services)

The *In Principle, In Practice* Initiative was designed to begin building a bridge between theory and practice, between what we currently know and what that implies for institutional practice and policy, and finally between what we are doing now and what we still need to do and learn in the years ahead to maximize the success of the museum community as part of a national and international learning infrastructure. The past decade has seen not only a growing desire for this information but a growing sense of urgency and need. The number of institutions has continued to grow exponentially and visitation at museums is at all-time highs; but against this background, many—if not most—institutions are feeling challenged by changes in the economic, social, and political landscape (Bradburne, 2004; Falk & Sheppard, 2006; Munley & Roberts, 2006; Silverman & O'Neill, 2004). The emphasis on learning experiences and growth in museum visitorship and visitation comes at a time when the mission of museums is shifting from a focus on collecting and preserving to one of educating the public. The late Stephen Weil (2002) described this as a movement from being *about something* to being *for somebody*. We believe this book, with its focus on visitor learning, will provide museum professionals of all types with an important tool to support their own learning and that of their institutions.

We have divided the book into four sections: How People Learn in Museums, Engaging Audiences in Meaningful Learning, Fostering a Learning-Centered Culture in Our Institutions, and Investigating Museum Learning in the Next Ten Years. Although the book was designed to be read cover to cover, like any edited volume we understand that not all readers will find all chapters equally useful or compelling. Thus, we encourage you to freely sample, selecting the section, or even the individual chapters, that best meet your immediate needs. Please use the book as a reference and if so inspired, perhaps even as a focus of discussion within your institution.

The first section of the book, How People Learn in Museums, provides a foundation for understanding the nature of learning in and from museums. Although not an exhaustive synthesis of what we currently understand about this extremely challenging subject, the five chapters in this section provide a fairly comprehensive overview of some of the most important findings from the past decade. The first and last chapters in this section provide a holistic view, emphasizing the complex, contextual nature of museum learning, and the need to address the longer-term impacts of museum experiences. Two other chapters tackle learning through the lens of arguably the two most important social groupings of museum visitors—families and school groups. Rounding out the section is a chapter that tackles the fundamental challenges of understanding the role that exhibitions play in supporting museum learning. As the five chapters in this section attest, considerable progress has been made over the past ten years in our understanding of museum learning, and readers will certainly come away with a better sense of what we currently do and do not yet understand about learning in and from museums.

Building upon the foundations of the first section, the second section, Engaging Audiences in Meaningful Learning, addresses some of the challenges facing museums in this time of economic, social, and political change. Each of the four chapters in this section challenges the current museum status quo. Chapters discuss developing more customized and personal experiences for visitors, addressing the issue of institutional authority and worldview in an increasingly multicultural society, the need to focus institutional mission upon socially relevant goals rather than narrow academic ones as has often been the case in the past, and the opportunities and risks of communicating controversial topics. Each of the chapters advocates for a readjustment and accommodation to changing realities—accommodations that involve significant challenges to the field but also represent major opportunities.

Appropriately, the third section of the book, Fostering a Learning-Centered Culture in Our Institutions, describes the structural changes 21st-century museums will need to undergo in order to engage their audiences in the meaningful learning discussed in the previous section. Each of the four chapters lays out initial guidelines for directly addressing institutional change—through reformulated business models; by focusing on the creation of a culture of learning within the organization; by systematically and consciously integrating learning theory, practice, research, and policy; and finally by fostering a culture of collaboration, both within and without the organization. Each of these approaches is challenging in its own right; however, each is fundamental to success in the knowledge age.

The last section of the book bookends the first and is appropriately titled Investigating Museum Learning in the Next Ten Years. The first section provided an overview of what we have learned about the nature of museum learning over the past decade; the final section lays out an agenda for the research we need to do in the next decade in order to lay a foundation for tomorrow's museum practice. As foreshadowed by earlier chapters, these four chapters advocate for new paradigms of research: paradigms that emphasize longer-term investigations, a better understanding of the social nature of museum experiences, and investigations that better accommodate the complexity of the museum experience.

As suggested earlier, we believe this book is an important marker in the development of the museum field, a time capsule of current best thinking from a representative cross-section of the museum community. We certainly do not assume that everyone will agree with all of the ideas postulated here. However, we purposely encouraged authors to be bold, so what we do believe is that we have assembled a provocative and thoughtful view of the future of museums. We will feel successful if the writings assembled here stimulate both reflection and debate and ultimately raise more questions than answers.

DEFINITIONS

The goal of the *In Principle, In Practice* Learning Innovation Initiative was to apply new understandings of free-choice learning to the development of visionary frameworks for improving the practice, evaluation, and future research efforts of the museum community. A major first step in this process was the bringing together of the varied parts of the community—diverse in role within the community (e.g., researcher, evaluator, practitioner, policymaker), diverse in venue and discipline (e.g., science center, art museum, aquarium, natural history museum, zoo, children's museum, history museum), and diverse in stage of career (e.g., participants included established leaders and emerging ones). In so doing, we hoped to provide a shared platform for understanding and communication while at the same time encouraging divergent thinking. Because of the diversity, though, it was essential that we share a common vocabulary and set of definitions. What follows are several key terms that we identified as important. We stipulated that all authors were to build from and use these definitions.

Museum—We use *museum* as a generic term to refer to all the various kinds of museum-like institutions, such as art, history, and natural history museums, science centers, zoos, aquariums, botanical gardens, and nature centers.

Exhibition/Exhibit—There is frequently confusion and a lack of specificity about the use of the terms *exhibition* and *exhibit*. For clarity, we utilize Serrell's (1996) definitions: an exhibit is an individual unit or element within a larger exhibition. In other words, collections of exhibits are combined to form exhibitions.

Learning—This is the most difficult definition of all. For purposes of this book, learning is defined as a personally and socially constructed mechanism for making meaning in the physical world. The definition is a broad one and includes changes in cognition, affect, attitudes, and behavior.

Informal—*Informal* is a term used to describe institutional settings other than (formal) classroom settings; for example, museums are informal settings. However, informal is *not* used to describe the nature of the learning occurring in the setting, since the fundamental processes of learning do not differ solely as a function of institutional setting.

Free Choice—This term describes the learning that occurs in settings in which the learner is largely choosing what, how, where, and with whom to learn. It is a generic term that captures the intrinsically motivated nature of most museum-based learning. However, it is important to note that although free-choice learning describes the learning of most casual visitors in museums, not all museum-based learning is free choice. For example, when children in school groups take field trips where there is a predefined lesson with limited or no choice or control over goals and activities, the learning is best described as compulsory.

Finally, it is with great pride that we launch this volume as the first in a new publication series—the *Learning Innovation Series*. The series is a joint effort of the Institute for Learning Innovation and AltaMira Press, devoted to the how, where, and whys of free-choice learning. Volume 1 in this new series—*In Principle, In Practice*—coincides with the twentieth anniversary of the Institute for Learning Innovation, a not-for-profit organization committed to understanding, supporting, and advocating for free-choice learning—learning that fulfills the lifelong human quest for knowledge, understanding, and personal fulfillment. Given our history, it is altogether fitting that this volume be devoted to museums, since for twenty years we have disseminated the results of our museum-based research, evaluation, and planning efforts through publications, presentations, and professional training. By providing museum-learning policymakers, researchers, and practitioners with up-to-date understandings of free-choice learning, we have been able to encourage a more encompassing and accurate understanding of how these institutions support an ever evolving learning society.

However, the *Learning Innovation Series* will not only be focused on museums. The series will provide the broader education community with comprehensive edited volumes on a wide variety of free-choice learning topics. For example, the second book in the series, to be published in 2008, will focus on free-choice learning and the environment, while future books are planned to focus on the areas of free-choice learning and media, libraries, public health, young children, and older adults. Individuals interested in developing new books in this series should contact the series editors, John Falk and Lynn Dierking, at falk@ilinet.org or dierking@ilinet.org.

John H. Falk,
Lynn D. Dierking,
Susan Foutz,
Institute for Learning Innovation

Acknowledgments

This book was made possible through the support and hard work of many individuals and organizations in the museum field. The editors are grateful to the National Science Foundation for funding *In Principle, In Practice: A Learning Innovation Initiative* (grant number ESI 0318868); John Wiley & Sons, Inc. for permitting contributors to update and republish articles previously published in *Science Education,* 88 (suppl.1), 2004; all those who participated in the November 2004 conference in Annapolis, MD; the contributors who collaborated on the chapters in this volume; the chapter reviewers; the American Association of Museums for hosting a session at the 2005 annual meeting on this topic; and those who contributed the photographs that open each section, the credits for which are listed below.

Looking at Orangutan Panels on the Tropics Trail
Courtesy of the Phoenix Zoo

Chumash Family Singers at the Aquarium of the Pacific
Courtesy of Jerry Schubel

Tinker Toy at the Exploratorium
Courtesy of Lily Rodriguez, © Exploratorium, www.exploratorium.edu

Institute for Learning Innovation data collectors at work
Courtesy of the Institute for Learning Innovation

I

HOW PEOPLE LEARN IN MUSEUMS

1

Toward an Improved Understanding of Learning From Museums: Filmmaking as Metaphor

John H. Falk

"Rhett . . . if you go, where shall I go? What shall I do?"—Scarlett

"Frankly, my dear, I don't give a damn."—Rhett
Gone With the Wind, 1939

Great films transport us to another place and time, and for an hour or two allow us to fully enter the minds and sometimes even the bodies of complete strangers. Great research should do no less, particularly if our desire is to apply those research findings to practice. I would argue that although significant progress has been made in understanding and describing learning from museums, researchers actually have yet to fully "enter the minds and bodies" of visitors to these institutions. In this chapter, I describe two perspectives that I believe could significantly move research in this more visitor-focused direction. To help make my points, I use filmmaking as a metaphor. Both filmmaking and investigations of learning represent attempts to encapsulate and describe a slice of life; each is an effort to tell a persuasive and coherent story within constraints. Hence, borrowing images from one realm can provide insights into the other.

PIECES AND WHOLES

Making a film requires bringing together many different components. Before the filming can even begin, there are scripts to write, locations to scout, actors to cast, a crew to hire, musical scores to arrange, costumes to design and assemble, and, of course, financing to organize, just to name the most obvious details. Once all of these particulars are in place, the director must then worry about choreographing the complexity of cameras and sound booms, stars and extras, set designers and special effects, as well as electricians and carpenters. The average film is composed of myriad elements—elements with which most filmgoers are simply unaware. Quite similarly, myriad variables underlie learning from museums, and recent efforts to understand learning from museums have focused on enumerating and measuring these many components.

Roughly a decade ago, Lynn Dierking and I began to describe, in broad-brush terms, the many interacting variables that comprise the museum experience and result in learning (Falk & Dierking, 1992, 1995). In large part, we were reacting to the prevailing behaviorist models of museum learning which postulated that, given the "right" stimulus (e.g., a well-executed exhibition and/or label), visitors would achieve the "right" response (i.e., learn what the museum intended them to learn). The commonly used constructs of "attracting power" and "holding power" derive from this line of thinking, as do notions of "time on task." Although behaviorism has fallen out of favor as a psychological model, its stamp on the field is still quite pervasive and many continue to utilize behaviorist-based constructs without realizing their origins.

Fortunately, concerns about such a simplistic model have increasingly taken a backseat as a growing cadre of investigators has sought to explore other dimensions of learning. Over the past decade, the list of variables that have been seriously investigated relative to visitor learning has grown considerably. As several of the articles in this volume attest, considerable attention has been directed toward that broad set of factors related to visitors' social and cultural contexts. A range of investigations now exists and shows that visitors to museums are strongly influenced by the interactions and collaborations they have with individuals within their own social group (e.g., Borun, Chambers, Dritsas, & Johnson, 1997; Schauble, Banks, Coates, Martin, & Sterling, 1996). A key focus of much of this research has been on the analysis of visitor conversations (see Leinhardt, Crowley, & Knutson, 2002). Research has also shown that the quality of interactions visitors have with individuals outside their own social group—for example, with museum explainers, guides, demonstrators, performers or even other visitor groups—can make a profound difference in visitor learning (Rosenthal & Blankman-Hetrick, 2002; Wolins, Jensen, & Ulzheimer, 1992). In addition, a number of individuals have postulated that cultural affinity also plays a role in how and what visitors to museums learn (e.g., Fienberg & Leinhardt, 2002; Paris & Mercer, 2002). Collectively, these investigations make us aware that learning from museums should be affected by within-group social interactions, by social interaction and facilitation from individuals outside the visitor's social group, and by the cultural values and beliefs visitors hold relative to culture and identity.

One of the striking things about museums is that they are physical settings unlike those most people encounter in their daily lives. Since museums are typically free-choice learning situations, the experience is generally voluntary, nonsequential, and highly influenced by the features of the setting (Falk & Dierking, 2000). As such, visitor learning has been shown to be strongly influenced by how successfully visitors are able to orient within the space (e.g., Evans, 1995; Kubota & Olstad, 1991); being able to confidently navigate within a complex three-dimensional environment turns out to be highly correlated with what and how much an individual learns. Similarly, intellectual navigation, as supported by quality advance organizers (Anderson & Lucas, 2001; Falk, 1997) has been shown to affect visitor learning from museums. Research has also shown that a myriad of architectural and design factors such as lighting, crowding, color, sound, and space subtly influence visitor learning (Evans, 1995; Hedge, 1995; Ogden, Lindburg, & Maple, 1993). And, of course, the particulars of exhibition design also influence visitor's behavior and learning (Allen, 1997; Bitgood, Serrell, & Thompson, 1994; Sandifer, 2003; Serrell, 1996).

Building upon constructivist theories of learning, the influences of prior knowledge and experience on museum learning have also been widely described and documented (Dierking & Pollock, 1998; Falk & Adelman, 2003; Roschelle, 1995), as has the role of prior interest (e.g., Adelman, Falk, & James, 2000; Csikszentmihalyi & Hermanson, 1995; Falk & Adelman, 2003). The exact nature of a visitor's motivations and expectations, or "agenda," for visiting a museum has also been shown to significantly influence the visitor's learning outcomes (e.g., Doering & Pekarik, 1996; Falk, Moussouri, & Coulson, 1998); and more recently investigators have appreciated that identity too plays a role in visitor learning (Ellenbogen, 2003; Falk, 2006; Rounds, 2006). Finally, it has been appreciated that the degree of choice and control over learning also affects visitor learning (e.g., Griffin, 1998; Lebeau, Gyamfi, Wizevich, & Koster, 2001). These studies have raised awareness for the fact that visitor learning is influenced by the realities of an individual's motivation for visiting and his/her identity. It also means that one should expect learning to be highly personal and strongly influenced by an individual's past knowledge, previous museum experiences, and personal interests. Also, one should expect learning to be influenced by an individual's desire to choose and control his/her own learning.

Arguably, a major strength of recent research on learning from museums has been the description and investigation of these many factors that appear to influence learning from museums. Theoretically, these factors, which directly and indirectly influence learning, number in the hundreds, if not thousands. Some are apparent; others are either not apparent or currently perceived to be unimportant. However, though we now understand the abundance of factors that influence learning in and from museums, we still tend to conduct research as if only one or two of these variables mattered.

It can be hard to be objective about your own discipline when you are immersed in it, so the film metaphor may be useful here. What would happen to a film if the director got fixated on only one piece of the process? To illustrate how this might happen, imagine that the producer made the set designer the director. She, not surprisingly, thinks that the sets and props are fundamental to the quality of the film—which, of course, they are. But given control of the camera, virtually all the footage seems to be of sets and props. The actors and actresses become almost superfluous. What does it matter who the actors are anyway? Once in the set, they'll all act the same. The producer decides this is ridiculous and turns the production over to the script writer—who, of course, is fixated on the dialogue. The new iteration of the film might as well be a radio talk show; scene and setting, lighting and mood, costume and makeup fall away as all attention is focused on the content of the conversations between actors. I could go on, but you probably get the point. A film is not just about sets and scenes, or about dialogue, makeup, costumes, casting, lighting, sound, music, film processing, editing, or any of the other numerous pieces that make up a film. No, a quality film from the viewer's perspective is the coming together of all these elements into a seamless whole. A skilled director understands the vital importance, and nuance, of all these elements;

rather than focusing on one or two, she weaves them all together, allowing the pieces to complement rather than distract from each other. Good research on learning from museums should do the same thing. All too frequently, it does not.

On the positive side, the theoretical frameworks that currently influence many museum-learning researchers—for example, the sociocultural perspective (see Leinhardt, Crowley, & Knutson, 2002; Martin, this book) and the constructivist-inspired Contextual Model of Learning (see Falk & Dierking, 2000)—advocate taking a holistic view. For example, the recent qualitative investigations of Leinhardt, Tittle, and Knutson (2002) and Fienberg and Leinhardt (2002) provide insights into the importance of the myriad details and complex interactions of individuals, settings, and situations. Both of these investigations highlight how variable visitor learning is and how strongly influenced it is by the idiosyncratic confluence of visitors' experiences prior to and during their museum visit. Martin Storksdieck and I (2005) recently examined the multivariability of visitor learning utilizing a more quantitative approach.

Our efforts to identify the factors that contributed to changes in visitor understanding revealed that, as cited above, variables such as prior knowledge, interest, motivation, choice, and control, within group social interaction, between group social interaction, orientation, advance organizers, architecture, and the quality and quantity of exhibits viewed all do significantly influence learning for at least some visitors. In other words, all of these factors were important; however, no single factor was capable of adequately explaining visitor learning outcomes across all of the nearly 200 visitors we studied. Even the most significant of these factors explained only about 9% of the variance in learning. The data better supported the idea that collections of factors, rather than individual factors, provided a reasonable explanation for the nature of visitor change in knowledge about biology, in this case. Perhaps most importantly, different suites of factors were found to have significantly affected different groups of visitors. The key to teasing out these effects, though, was in being able to meaningfully segment visitor groups. Traditional demographic categories like age, race/ethnicity, and even social group and educational attainment proved of limited usefulness in this respect. By contrast, grouping visitors as a function of their prior knowledge of biology, motivation for visiting the museum, and prior interest in biology were useful. For example, for visitors with the most limited entering knowledge of biology, exhibit quality emerged as a highly significant contributor to learning; orientation to the exhibition space had no significant effect on this group. However, exhibit quality was not an important factor for visitors with greater knowledge of biology; for this group, orientation to the exhibition was a significant contributing factor to their learning.

The results of this investigation showed that learning depended individually and collectively upon who the visitor was, what he knew, why he came, and what he actually saw and did. As in filmmaking, creating a compelling story about the nature of learning in this setting was significantly enhanced by an ability to accommodate multiple variables; focusing on only one or two variables would have yielded a limited view of what actually happened. That said, even when all of our variables were considered, they did not provide a complete picture; at best, we could only explain about half of the variance.

Some of the deficit was undoubtedly caused by deficiencies in measurement. Although we attempted to measure each of our variables utilizing the best methods we could borrow or invent, all were exceedingly complex in their own right and only partially amenable to fully valid and reliable measurement. Beyond measurement errors, though, some of the unexplained variance was caused by the very real impact of seemingly random events. The "initial condition" of visitors (e.g., prior knowledge, motivation, interest, and social group and role) were very important contributors to visitor learning, but these conditions changed over time through interactions with both predictable (e.g., presence of advance organizers and the nature of exhibits visitors interacted with) and unpredictable events. Unpredictable events included, for example, a large group of visitors clustered around an introductory exhibit panel, causing a visitor to skip reading the advance organizer for the exhibition; an accompanying child needing to go to the toilet, which truncated the visit; and a text panel including information the visitor's partner just happened to have read about the previous day. Unpredictable events such as these strongly influenced not only which factors came into play as important but also modulated the relative amplitude of impact those factors had on learning; unpredictable events influenced both the quality and quantity of learning. Although it is frustrating to appreciate the role of random events, they are a reality in learning settings such as museums and need to be considered.

Thus, envisioning the museum experience as a series of cascading events acting upon a set of "initial conditions" defined by the characteristics of the visitor and her social group provides one way to begin to conceptualize and assemble the multiple pieces of the museum experience into a conceptual whole. However, this model, too, appears to have limitations. The model, or at least how we and most others have interpreted and implemented it, only partially accommodates the temporal and physical context of learning. The model as described above is deficient in scope and scale; it is a whole but an insufficient whole. It places too great an emphasis on what is happening within the museum. Although both the sociocultural model and Contextual Model of Learning argue for the need to consider learning within a broad temporal and physical scope and scale, relatively few researchers have actually approached their investigations from this perspective. As it turns out, an enlarged scope and scale are not just abstract niceties.

SCOPE AND SCALE

Just as it is possible to make a film that only captures the life of one person for one or two hours within a single setting, so too is it possible to dissect and study an individual's learning from a museum within the delimited physical scope and time scale of the actual museum visit. In both cases, the results are likely to yield a very close-up view of a moment in time, a view that makes it hard to know how to place that moment in appropriate context. For example, the exquisite 1984 film by Bertrand Tavernier, *A Sunday in the Country*, richly portrays the events of one Sunday afternoon in pre-WWI France when the family of an aged painter comes for a visit. There are no flashbacks and no fast-forwards; just the events of a couple of hours, on one day, in the country. When the film ends you want to know, who was this painter? Why were his children feeling the way they did about him? What happened next? Although the film is cinematically beautiful and the acting superb, the limited context makes it difficult, if not impossible, to actually understand what was happening on that lovely afternoon in the country.

In similar fashion, for many years it has been standard practice to conduct investigations of museum learning exclusively within the temporal and physical space of the museum. Despite being feasible, practical, and outwardly reasonable to investigate museum learning in this way, an increasing array of data suggests that doing so yields insufficient context to meaningfully talk about what visitors actually learn.

One of the important take-away messages from the Falk and Storksdieck research discussed above is that it reinforces that understanding learning in general, and learning from museums in particular, requires knowledge of the individual, which in turn implies the need for the much longer view advocated by Falk and Dierking (1992, 2000). Each person is unique, a product of his or her personal history and development. Thus, from this perspective, in order to adequately understand an individual's learning requires knowing something about the person's prior experiences, prior learning, stage of intellectual, physical, and emotional development, cultural and social history, and interests and expectations for learning in this particular situation. I have suggested that all of these coalesce into a particular identity and yield a specific "motivation" for learning situated in that moment and that circumstance (cf. Falk, 2006).

Motivation is a fundamental but historically underappreciated part of learning (Csikszentmihalyi & Hermanson, 1995; Deci & Ryan, 1985; Dweck, 1989; McCombs, 1991, 1996). This is perhaps not surprising when you consider that for years the majority of learning research was conducted in either laboratory or school settings—settings in which the motivations of the learner were considered of little importance (Schoenfeld, 1999). In free-choice learning situations, motivation emerges as central, influencing not only the "what" of learning but also the "why" and "how." Learners in free-choice learning situations, as exemplified by casual visitors to a museum, self-select what to learn about based upon their interests and prior experiences, why to learn based upon their motivations and expectations relative to the setting, and how to learn based upon their learning styles, development, and social/cultural preferences. What is learned today thus depends greatly upon an individual's motivations and identity, which in turn are determined by what was learned yesterday. In this view, learning is a whole, not a part—a whole that can only be understood by trying to situate any given learning experience within the larger framework of a person's total life. In effect, to investigate any given learning situation requires panning the camera back in time and space in order to see how this particular situation fits within the larger picture of a person's life. Filmmakers use a variety of devices to accommodate this need. A common strategy is to insert flashbacks into the film, or sometimes voice-over narration that provides the viewer with sufficient context to know why an actor behaves in the way he does when confronted by some event. The magic of film is that it also allows the filmmaker to fast-forward into the future; this device also gives the viewer context by allowing him to see how a seemingly inconsequential event in the present will end up being important in the future (Alfred Hitchcock was the master of this technique). As seems to be better understood by filmmakers than museum-learning researchers, understanding the meaning of the present requires knowing something about both the past and the future.

The "future" part of this statement has only recently begun to be appreciated by museum-learning researchers. Our brains build new understandings through a continuous process of perceiving new patterns, images, and ideas and accommodating them within existing structures (Damasio, 1994; Edelman, 1987). This is an active, ongoing process—so ongoing, that recent evidence suggests that memories are only partially laid down at the time of an event. The whole process occurs over hours, days, and sometimes even weeks (Nader, 2003; Schacter, 1999). Hence, asking someone like a museum visitor to tell you what she remembers about her experience immediately following the experience yields, at best, limited information, and at worst, potentially misleading information. It is not that individuals will have no memories of the experience; they will have memories. Importantly, though, the memories they have at that instant are as likely as not to be different than the memories they will have a few days, weeks, months, or years later. This fact turns out to be extremely important, because just as there is no single right way to learn things, there is no single place or even moment in which learning occurs. Several recent museum-learning investigations reinforce that learning happens continuously, deriving from many different sources, and in many different ways (see Anderson, Storksdieck, & Spock, chapter 14 of this book, for a review). Learning is only very rarely of the "eureka" variety, blinding insights that allow us to connect previously unknown and unrelated ideas. Most learning involves the continuous piecing together of new bits of information into existing ways of thinking, what Piaget (1952) called assimilation. Each new experience deepens and broadens our understanding, each becomes part of a larger whole (Bransford, Brown, &

Cocking, 1999)—what Barab & Kirshner (2001) call "learning in the making." Thus, in order to understand an individual's learning requires at least some rudimentary sense of that person's learning history—past and future. But in this lies a paradox.

As any historian will tell you, it is almost impossible to "objectively" analyze a current event. There is insufficient perspective from which to sort out what is important and what is not. Only after a suitable lapse of time is it possible to make these judgments. So, how are we to accurately measure learning in the present if we cannot objectively judge all of the events, past and future, that combine to define that event? The simple answer is we cannot. However, what we can do is mindfully sample events in such a way that we gain as accurate a view as possible. Some clarity can be found by once again using a moviemaking metaphor.

Making a credible film of an event is not just about pointing a camera and shooting, as was painfully evident to anyone who watched some of Andy Warhol's early movies. In order to capture both the overall story as well as the richness of an event requires considerable planning and thought since the camera cannot simultaneously capture all aspects of a scene; at different times, some angles and views provide richer, more useful information than others. Each lens of the camera has discrete capabilities; some provide close-ups while others provide wide-angle views. Every filmmaker must make choices, what to include and what to leave out. So it is with investigating something as complex and long term as learning. Selecting a wide-angle lens during investigations enables us to view the experience over longer periods of time. The longer the time frame, the more of the context of learning is captured. Greater context affords greater insights into how a particular learning experience is situated within an individual's life and thus greater ability to explain both why something was learned as well as what was learned. However, just as a wide-angle camera lens provides a large overview but misses the details, potentially so can longitudinal studies of learning. The investigator may capture the gestalt but lose the specifics. By contrast, using a telephoto lens allows the filmmaker to zoom in on the details of the scene, capturing the richness of the costumes or the sudden but subtle change in an actor's expression. However, the tighter the shot, the more limited the context—true in both filmmaking and learning research. Recording and analyzing a conversation between two individuals as they stand before an exhibit can provide a very fine-grained understanding of the learning process, but in the absence of a larger context may provide little actual insight into the motivations behind the conversation or what the ultimate effect of this conversation might be. In theory, then, just like a good film, a good learning study should provide an appropriate mix of close-ups and panoramas; it should capture sufficient detail to explain the processes that the individual is engaged in, while at the same time capturing longitudinal aspects—flashbacks and fast-forwards—that allow the learning experience to be situated within a larger context in order to make sense of why and what learning occurred.

Appreciation of the importance of scope and scale amongst investigators of museum learning has come slowly and most efforts to accommodate these ideas are still quite crude. Although some in the field still assume that learning from museums means the addition of "new" knowledge, there is a growing awareness that much of the learning afforded by museums is in the area of consolidation and reinforcement of previous understandings and perspectives (see Anderson, 1999b; Ellenbogen, 2003; Falk & Dierking, 2000); individuals are much more likely to describe the outcomes of their museum experiences as strengthening rather than changing their existing knowledge structures. For example, a random telephone interview of Los Angeles residents showed that 95% of those individuals reporting previously visiting the California Museum of Science and Industry believed that their museum experience had strengthened or extended their knowledge of some area of science while, on a separate question, only 66% felt that their museum experience had changed their understanding, attitudes, and/or behaviors (Falk, Brooks, & Amin, 2001).

Over the past decade, a growing number of museum-learning investigators have begun to incorporate longer time frames into their research. Recent longitudinal studies have shown that the learning that results from a museum experience does change over time, and not always just by declining (Adelman, Falk, & James, 2000; Anderson, 1999b; Bielick & Karns, 1998; Ellenbogen, 2002, 2003; Falk, Scott, Dierking, Rennie, & Cohen Jones, 2004; Luke, Cohen Jones, Dierking, Adams, & Falk, 2002; Medved, 1998). What these studies reinforce is that long-term outcomes are often not predictable from short-term outcomes. Thus, in order to understand visitors' museum learning requires understanding visitors across three "time" periods: (1) visitor premuseum history—in particular, visitors' prior knowledge, interest, experience, expectations, and motivations; (2) visitor in-museum experiences—in particular, the actual experiences visitors have with specific exhibitions, their social interactions both within and outside their own social group, as well as characteristics of the physical setting such as crowdedness and presence or absence of advance organizers; and (3) visitor postmuseum history—in particular, the types of reinforcing experiences visitors have, such as postvisit conversations, reading, and television watching.

CONCLUSIONS

A major allure of investigating learning in and from museums is that it provides a setting in which to investigate learning that more accurately mirrors the way most people learn most of the time. Environments developed to support real-world learning such as museums are not mere backdrops for supporting the transmission of knowledge, they are what Barab and Kirshner (2001) call "dynamical learning environments." As such, these settings are always multidimensional, dynamic, and complex (see Brown, 1992; Cobb, Confrey, diSessa, Lehrer, & Schauble, 2003; Collins, 1999); our efforts to understand learning from museums needs to be no less so. Unquestionably, our understanding of learning from museums has made great strides in the past decade or so, much of it attributable to focused investigations that have allowed investigators to hone in on the specific variables and learning outcomes that result from visitor interaction with museum exhibits. In this chapter, I have attempted to make the case that in the next 10 to 15 years we will need to design our investigations with a new level of sophistication in order to accommodate the true complexity of learning from museums.

I have argued here that understanding learning in and from museums requires the simultaneous investigation of multiple variables. Clearly, a major message is that just as we now know that many factors must be considered when designing museum-learning experiences, we should also now know that many variables must be accounted for when attempting to understand the learning outcomes that result from these experiences. However, the real take-away message is that simple, reductionist approaches to understanding learning from museums will simply not suffice.

Only by appreciating and accounting for the full complexities of the museum experience will useful understandings of learning from museums emerge; we need to understand both the pieces and the whole.

Clearly, the solution is to be more holistic in our research; however, in the process, we must also not forget that the whole is larger than we have traditionally assumed. Learning from museums cannot be easily delimited in space and time. Our investigations must be designed to accommodate the broader physical and temporal scope and scale in which learning from museums actually occurs.

The promise of the research cited in this chapter and other chapters in this book, along with similar studies still in the implementation or planning stages, is that an ever-more refined model of learning from museums seems possible. Even though it is clear that learning from museums is complex, it is equally clear that it is a topic amenable to analysis and description. As with all metaphors, the filmmaking metaphor eventually breaks down. Our efforts to understand the real-time, unknown experiences of visitors is nothing like creating a typical Hollywood blockbuster; we don't have the luxury, or desire, to invent whatever scenes and stories we want. However, one type of filmmaking does come close to our situation: documentary filmmaking.

I would assert that there is much we can learn from the skilled documentary filmmaker, who like a good researcher is trying to determine how to capture "reality" often in "real time." Although the filmmaker may shoot footage in "real time" as events are unfolding, typically he is also shooting additional footage—interviews with people who knew the protagonists prior to the event, interviews with people after the event took place, footage of events in other places and times that provide context for the event in question—all of which are ultimately spliced together to make a coherent story. The experienced documentary filmmaker realizes that it is difficult, if not impossible, to always know beforehand what footage will be needed in the final cut. Although he may have a basic plan in his head prior to a shoot, he comes prepared to shoot more film than he will ultimately use; he comes with a range of lenses so that he can take both long shots and close-ups as the situation demands.

Thinking along these lines need not change the basic guiding principles of any good investigation (see Shavelson, Phillips, Towne, & Feuer, 2003). What I would argue should change, though, is how we conceptualize and implement these basic research principles. Providing a reasonable, accurate, and—equally important—compelling story does not just happen. Like good documentary filmmaking, quality learning research requires collecting a wealth of data and sifting through it and then reassembling the pieces into a multilayered, compelling, accurate, but still comprehensible story—a story of real people, living real lives. I would like to believe we are on the threshold of that ability now. Without a doubt, approaching research in the way advocated in this chapter is much more costly and time consuming than our current approaches. However, if this is what it takes to make sense of visitor learning in and from museums, then adjust the lights and let the cameras roll. It is only in doing so that we will conduct research that actually can usefully inform practice.

ENDNOTES

An earlier version of this chapter was published in *Science Education*, 88 (suppl.1): S34–S47, 2004, © Wiley Periodicals, Inc. Used with permission of John Wiley & Sons, Inc.

2

Family Learning in Museums: Perspectives on a Decade of Research

Kirsten M. Ellenbogen, Jessica J. Luke, and Lynn D. Dierking

In the spirit of the *In Principle, In Practice* initiative, this chapter discusses how research focused on family learning in and from museums has progressed over the past decade and shares some perspectives from the field regarding its usefulness for practice. Although families were clearly a dominant audience in museums, there were not significant numbers of researchers focusing on family learning until the mid to late 1970s. By the late 1980s, there was an extensive body of literature that established the importance of family learning in and from museums (see Astor-Jack, Whaley, Dierking, Perry, & Garibay in chapter 15 of this volume).

These early studies demonstrated the significance of families as a focus of museum research, identifying them as a major audience and unique learning group of mixed ages and backgrounds bound together by a complex shared system of past experiences, beliefs, and values. They also established the complex nature of family interactions, highlighting the ways in which family members interact and learn together, and providing evidence that families bring an extensive array of personal and cooperative learning strategies to their experiences in museums. Much of this research was descriptive in nature, depicting family conversations, the roles adults and children assume, the influence of specific conversational rules on physical behavior, gender differences in parent-child interactions, and the similarities and differences between the conversations of families and all-adult groups.

Ten years later, we are reaping the benefits of influential conferences (e.g., Falk & Dierking, 1995; Paris, 2002), benchmark political reports (e.g., the American Association of Museums [AAM], 1992), the realization of a long anticipated "learning society" (e.g., Falk & Dierking, 2002; Hutchins, 1970; Malchup, 1962), and participation from researchers that represent increasingly diverse research backgrounds. We face a pivotal point as a community, with opportunities to offer insights and recommendations to those working with families in museums.

We highlight three aspects of this new research to support our assertion that important progress has been made that holds great potential for influencing practice in the area: (1) shifting theoretical perspectives that signal shared language, beliefs, values, understandings, and assumptions about what constitutes family learning; (2) realigning methodologies that are driven by underlying disciplinary assumptions about how research in this arena is best conducted, what questions should be addressed, and criteria for valid and reliable evidence; and, (3) resituating research foci to ensure that the family is central to learning, reflecting a more holistic understanding of the family as an educational institution within the larger learning infrastructure. We also explore connections between the theory and practice of family learning in museums, featuring reflections written by three museum professionals, highlighting their perceptions of what the progress in research has meant for developing meaningful museum experiences for families.

SHARED LANGUAGE, BELIEFS, AND VALUES OF THE FIELD

A major shift in the past decade of family learning research in museums stems from the use of sociocultural theory in learning research and a recognition of its suitability for museum research. Such a perspective frames learning in and from museums as socially and culturally constructed through people's actions within a specific community of practice. A community of practice shares a set of values, vocabulary, understandings, and assumptions (Wenger, 1998). A person's actions and interactions are interpreted by members of his group (Green & Meyer, 1991; Gumperz, 1986), allowing him to construct meanings specific to the group through his conversations (Geertz, 1983; Green & Dixon, 1993). This shift may seem subtle, but the implications of embracing a sociocultural perspective are significant. Studies from this perspective focus not just on the immediate experiences of the family group in the museum but more broadly on the ways in which the family group is situated within the larger social and cultural context. It necessitates understanding the shared meanings, processes, artifacts, symbols, and identities that families construct as they participate within a specific community of practice, more fully revealing the nature of the learning processes and products.

Although there has been research on families' conversations in museums for more than a decade (see Astor-Jack, Whaley, Dierking, Perry, & Garibay in chapter 15 of this volume), this research more generally described how families interact and talk about topics presented in exhibitions and programs. Family members talk about what they know from previous experiences, discussing what they see, hear, read, and do in relation to their family experiences and memories. This research also demonstrated that these discussions provide opportunities for family members to reinforce past experiences and family history, and to develop shared understandings.

The recent dominance of sociocultural perspectives in museum research builds on the early studies of family conversations and enables current investigators to pursue a depth of understanding about family conversations and their role in identity building and other social and cultural aspects not understood previously. In particular, these perspectives have afforded opportunities for more in-depth investigation of the meaning and uses of conversations (see Leinhardt, Crowley, & Knutson, 2002). Recent studies of families' conversations emphasize the processes families engage in to construct meaning and build identity, and the role of the museum experience in the family's larger social and cultural context (Ash, 2003; Crowley et al., 2001; Ellenbogen, 2002).

One study conducted as part of the Family Learning Initiative, a systemic research effort at The Children's Museum (TCM) of Indianapolis, illustrates the recent theoretical focus on the family group and the shared meanings they construct through participation in the larger social and cultural context (Luke, Cohen Jones, Dierking, Adams, & Falk, 2002). The study examined the long-term impact of two youth-based museum programs on young adults and their families. Findings demonstrated that these programs influenced participants' attitudes, interests, and awareness. However, the focus was not just on the impact at the level of the individual. The study employed a social systems approach to understand the long-term impact of participation in these two TCM program experiences within a larger sociocultural and developmental context. Within this expanded approach, findings demonstrated that these programs influenced family dynamics, giving young adults the opportunity to explore new roles within their family, gain new perspectives and identities within the family system, and learn new things about family members. There was evidence that interests that young adults developed within the program were carried over into the family context, resulting in shared family interests.

Programs also influenced young adults' contributions and connections to the larger sociocultural community, fostering a tolerance of other people and cultures, and cultivating a sense of civic responsibility. These results suggest that a research perspective informed by sociocultural theory highlights not only the learning of the individual but also learning at the level of social groups, such as family and community. Such shifts in theoretical perspectives also allow researchers to understand learning that is broader than content knowledge, such as learning to communicate with others, learning new things about family members and the family dynamic, and learning a sense of civic responsibility. These findings are building a shared language and set of beliefs and values in the field and are evidence of the progress that has been made in family learning research in museums over the last decade.

REALIGNING METHODOLOGIES

In the past ten years, researchers have attempted to more broadly and deeply investigate the nature of family learning in and from museums utilizing a variety of methodologies. Some of these methodologies are borrowed directly from other disciplines; others are unique improvements upon what has been developed previously. This change in methodologies is a natural realignment that accommodates shifts in assumptions about how research is best conducted, what questions should be addressed, and what is agreed upon by the community as valid and reliable evidence of family learning.

Researchers are adopting diverse methodologies that include discourse analysis; video and audio recording of moment-by-moment interactions; pre-, post-, and post-post-interviewing; journaling; and talk-aloud cued visits; and are providing family members with cameras as a documentation and meaning-making tool, all in an attempt to better understand and document the role that museums play in families' lives. These innovations are due, in part, to the strength of recent funding catalysts in the field that have supported conferences and research collaborations. In addition, museum researchers have become more aware of cutting-edge research on family learning in other fields of research (e.g., Moussouri, 1997). Innovations in methodologies have also arisen organically as researchers draw upon their diverse training backgrounds and adapt preexisting methodologies to family learning research in museums.

An example of the ways in which new methodologies have informed our understanding of family learning in museums is a multicity, multimuseum study focused on exhibit characteristics, family behaviors, and family conversations designed to identify the characteristics of exhibits that encourage family interaction (Borun et al., 1998). Multiple testing of the audiotaping methods used to record families' conversations resulted in a measure of the correlation between families' physical and verbal interactions and their learning, as measured through an interview process after families left an exhibition. Findings suggest that families engage in three levels of discourse: identifying, describing, and interpreting and applying. Findings from the second phase of the project concluded that families are learning from museums, but parents' mediating strategies are sometimes not the most effective for facilitating science learning.

Another example is Personal Meaning Mapping (PMM) (Falk, Moussouri, & Coulson, 1998), which is based on concept-mapping techniques and allows participants to respond to a concept, topic, or experience in their own ways. Researchers then use a structured system to code responses and conduct a quantitative analysis of the qualitative data. In a study investigating families' understandings of evolution (Luke, O'Mara, & Dierking, 1999), groups were encouraged to work together to write (or draw) anything that came to mind when they thought about a topic prompt. As they wrote and discussed their ideas, researchers recorded these negotiations to capture the family dynamic at work. Researchers used families' responses as prompts for a follow-up interview with the group. In this way, PMM allowed researchers to start with families' individual and collective perceptions of the topic and pursue their understandings in greater depth using their own language and terminology.

The realignment of methodologies has also pointed to the need to extend research over time. Some of the earliest museum-learning research (Robinson, 1928), as well as recent calls for such research designs (Falk & Dierking, 1995), note the importance of understanding what people do before and after their museum visit. Most commonly, an extended approach is taken by conducting a follow-up interview with families weeks, months, and even years after the museum visit. Researchers have repeatedly shown that many of the conversations that begin in the museum continue once families are back in the home (see Astor-Jack, Whaley, Dierking, Perry, & Garibay in chapter 15 of this volume). In addition, families are able to describe specific exhibitions and program elements without prompting, indicating the general durability of the museum experience.

For example, a study of the impact of a genetics exhibition included follow-up telephone interviews with a sample of families who had visited the exhibition approximately six weeks earlier (Luke, Coles, & Falk, 1998). Families reported that they talked about the exhibition in the car on the way home, over dinner in the following weeks, or while engaged in some other activity, and described connections between the content in the exhibition and other circumstances or phenomena in their lives.

Ethnographic case studies that involved long-term relationships with a set group of families who visited museums frequently, allowing repeated observations and interviews before, during, and after museum visits (Ellenbogen, 2002, 2003), suggested that conversational connections among museum experiences and real-world contexts are frequent and pervasive.

A MORE HOLISTIC UNDERSTANDING OF THE FAMILY AS A LEARNING INSTITUTION

Shifting research to a new theoretical perspective and realigning methodologies are critical steps in moving family learning research in museums forward. The empirical work emerging in the past decade focuses on families and their conversations, and examines learning beyond the walls of the museum. However, this is not enough; with this approach we are in danger of only superficially documenting the nature of family learning. At a broader level, we must understand family culture as dominant, resituating the focal point of what we study from the museum agenda to the family agenda, understanding the family as a learning institution within the larger learning infrastructure they inhabit, and the culture in which they function.

The term "family agenda" (in contrast to "museum agenda") highlights the need for museums to be aware not only of themselves as resources for learning but also of what each family might choose to take from the experience and each family's role in museum learning (Falk, Balling, & Liversidge, 1985; Hilke, 1987). This inverted point of view highlights the reality that families bring resources with them that in turn influence the ways they interpret their museum experience. The term "family agenda" emphasizes the need to recognize and accommodate the resources families bring to the museum in order to create a successful family learning context. Researchers have argued that not only do families have agendas for their museum visits but that these agendas directly influence the impact of the museum experience.

A radical interpretation of the notion of family agenda suggests that research that strives to fully understand how families learn in and from museums should situate itself within the educational institution of the family. Once the family is the starting point, the museum becomes one of many learning resources that the family uses. From this point of view, the position of the museum within the learning infrastructure (Falk & Dierking, 2002; Lewenstein, 2001; St. John & Perry, 1996) depends on the culture of the family—a more realistic reflection of real-world experiences.

When the focus of the research is resituated to a family-centric perspective, identity building becomes more significant. Identity is socially situated with respect to people's ongoing membership in specific communities of practice (Wenger, 1998). In Wenger's view, people's development of identities is an integral aspect of their participation in the practices of a community. Learners are members of a community to the extent that they have learned normative ways of thinking and acting that have been established by that particular community. Much of this learning is implicit and involves assumptions and rules about particular ways of speaking and participating in that community.

Consequently, identities are not fixed or static but are in a constant process of formation (Holland, Lachicotte, Skinner, & Cain, 1998; Kress, 1995; Wenger, 1998). People construct and contest identities through what they do and say. The development and negotiation of identity is influenced not just by community but also by the organizations and institutions of the community. The museum, like all educational institutions (Bruner, 1996), can be seen as a place of enculturation (Pearce, 1994). Enculturation involves developing identity as a part of a community, and the museum is one of the organizations that influence this activity. From this perspective (e.g., Ivanova, 2003), we can begin to examine museums and other institutions in the learning infrastructure as places for building and affirming identity.

Consider the findings of a study investigating a partnership among the Franklin Institute Science Museum and three inner-city Philadelphia schools designed to cultivate collaboration between teachers and parents in support of elementary children's science learning (Luke, Bronnenkant, & Dierking, 2003). Interviews with parents revealed that after only one year, participating families integrated elements of the program into their family life. In addition, parents felt that the program activities changed their family learning dynamic, giving them an opportunity to step outside their traditional role as homework "dictator," and to instead work alongside their children to do activities and feel comfortable not having all the answers. Similarly, in the ethnographic case studies of family learning described earlier (Ellenbogen, 2003), families were found to use supporting interaction strategies across learning environments. Most interestingly, the families, over time, used seemingly unrelated interactions in museums and at home to construct their family identity.

Studying the families' interactions across multiple learning environments provided a needed lens for understanding the complex motivations underlying the families' practices in the museum. Findings suggest that museums can be tools for enculturation that families use to establish and negotiate their identity. In other words, the museum is context, not content. The subject of investigation thus becomes the ways in which family members come to make meaning through their interactions in a setting, rather than the setting dictating those interactions and any subsequent learning. The families themselves function as learning institutions, drawing upon museums as one of many tools they have to build family identity. This focus on the family as central to the meaning-making process is additional evidence for how the development of family learning research in museums over the past decade has led to a resituating of research foci, with the hope of ultimately providing useful insights that can shape exhibitions and programs that enable families to shape and affirm *their* identities.

CONCLUSIONS

The past decade has witnessed a marked increase in our understanding of how, what, when, and where families learn in and from museums. Three aspects of recent research provide evidence for the importance and relevance of this field and the progress that has been made. First, shifts in our theoretical perspectives, with a new focus on the family as a type of social group and the shared meanings they construct, have resulted in a shared vocabulary and set of beliefs, values, and understandings about what constitutes family learning in museums. Although classification and semantic arguments, such as the accuracy of indicators of learning or the definition of the term "family," still occur, they are a natural outcome of an emerging community of research that is vetted and debated by a group of researchers with shared interests but different backgrounds and approaches.

Second, a realignment of our methodologies, including the adoption of new and alternative strategies for documenting the role of the museum in family life, has united many of us in the pursuit of similar questions. Although there is variety in what constitutes valid and reliable evidence for family learning, shared values and ideas about how to best investigate it are emerging within the community.

Third, ensuring that the family is the central focus of our research has resulted in an understanding of the family as a learning institution, utilizing the free-choice learning resources of an extensive learning infrastructure in order to construct and affirm individual and collective identities. The past decade of research has produced significant evidence of the ways in which the museum provides a context for such identity building. There is growing agreement that family learning is best examined from the perspective of the family and the larger learning infrastructure.

However, the real evidence that we are making progress in this arena will be a discernable difference in the next decade of family learning research and practice in museums, perhaps even an influence on research and practice in the broader museum community. A similar exercise of reflecting upon the past decade of research ten years from now should result in definitive evidence for changes in research and practice. What would constitute definitive evidence for these changes?

First, we would see a well-established, shared vocabulary and set of beliefs, values, and understandings about what constitutes family learning.

Although healthy debate would still be a natural part of this community of research and practice, the nature of the arguments would be different. The community would no longer debate the definition of the term "family" or argue about indicators of learning; instead, they would engage in conversations about the nuances and subtleties of family learning in museums and investigate questions such as: How do families appropriate the museum experience and use it to support their own needs? How do the experiences families have in museums relate to their other learning experiences? Practitioners would find the research useful to their practice and be able to utilize it to improve their abilities to support family learning and identity building. The community of researchers and practitioners would be drawn together by shared interests, with respectful and productive ways of talking together and debating these issues.

Second, we would see new and refined strategies for facilitating and documenting the role of the museum in family life. Although there would still be some variety in what constitutes valid and reliable evidence of family learning, there would be accepted ways of framing, facilitating, and investigating family learning within the community. There also would be some consistency in the important questions to be resolved in the field and the methodologies used to investigate these questions. Methodologies that are sensitive to social interaction, such as discourse analysis, video and audio recording of moment-by-moment interactions, journaling, and talk-aloud approaches, would be well-established, commonly used approaches. These approaches would be integrated into practice in meaningful ways also.

Last, it would be understood that a meaningful examination of the nature of family learning in the museum requires a central focus on the family. The family would be seen as a learning institution that utilizes the learning resources of an extensive learning infrastructure to build its individual and collective identity. No one would dispute that family learning is best examined and supported from the perspective of the family and the larger learning infrastructure.

A key indicator of the progress made in family learning research over the past decade is the extent to which this empirical work informs the work of practitioners. What has family learning research meant to exhibition and program developers? How do they use this research to influence their work? Here are the reflections of three practitioners; we hope that their stories will provide useful information to other practitioners eager to integrate more research-based approaches into their activities and to researchers regarding investigations that are most helpful to furthering practice.

Cathleen Donnelly, Exhibit Developer, and Leslie Power, Director of School Services and Family Programs, The Children's Museum of Indianapolis (TCM)

Both of us became a Family Learning Leader by participating in a comprehensive professional development program designed to build internal capacity about family learning among key management staff, representing educators and exhibit developers and also membership, curatorial, visitor, and volunteer services. The goal was to build our capacity in family learning and to help us develop skills to inspire our staff, and ultimately to mentor and coach them in the principles of family learning. The training was rigorous, including four two-day seminars, advanced readings and assignments, field trips, and a special project designed to integrate family learning research into the specific day-to-day tasks of the leader. Here are two examples of how this experience has transformed our work at TCM.

Exhibitions

Family learning research has changed the way we develop exhibitions at TCM. Previously, we developed, tested, and evaluated our exhibitions based on their appeal to children. During the exhibit development process, we now consider the family a learning unit, and based on family learning research, we:

- Test exhibition concepts with children *and* adults in focus groups made up of family members of a variety of ages.
- Design interactives for family accessibility, collaboration, and conversation. If we do not observe families playing and talking together, then the interactive is redesigned and retested. We developed a family learning matrix that incorporates a variety of family learning characteristics and is used as a tool by the core team as interactives are designed and tested.
- Select objects based on what we hear and see families engaged with as they view informal artifact displays and interact with curators. If

we can determine which objects spark interest, encourage conversation, and help families make personal connections, we know we will be more successful in attracting and holding their attention within an exhibition.

- Write labels to be read aloud—either between a parent and a child or a sibling and a sibling—and content is geared toward helping families make personal connections. Font style and point size are designed for a wide age range—from beginning readers to grandparents.

School Services and Family Programs

Family learning research has directly influenced my work developing family programs and school field-trip programs at TCM. Based on family learning research and the museum's mission of reaching children *and* adults, the education department decided to eliminate all school-age programming, including summer day camps, and to replace this programming with programs specifically designed for families. The challenge and goal of the family programs is to provide a range of family learning experiences that foster family interactions and meet the developmental needs of a variety of ages. We now offer family trips, family nights at the museum, family nights at local schools, and programs for home-schoolers.

The most exciting aspect of my work currently, which is a direct result of the Family Learning Initiative we have been engaged with in collaboration with the Institute for Learning Innovation, is the addition of a public scholar position at Indiana University-Purdue University Indianapolis (IUPUI). IUPUI and TCM collaboratively hired a "public scholar of museums, families, and learning" in the fall of 2005. This shared position requires a great deal of collaboration between IUPUI's Museum Studies program and School of Education and TCM. One area of future study will investigate the role of chaperones with small groups of students, determining whether current research on family interactions in museums can be extended to such groups.

Dale McCreedy, Director, Gender and Family Learning Programs, Franklin Institute Science Museum

In considering how family learning research has informed my own practice, I needed to think back across my 18 years in the field. Although it is difficult to imagine my efforts within the field not being informed in some way by research, the question that arises for me is "Which body of research has been most influential?"

In the early days of my work, gender research was most salient. This included the work of Sue Rosser, the American Association of University Women (AAUW), and many studies acknowledging the invisibility of females and minorities in texts, as role models, and within museum exhibitions. "My Daughter the Scientist," an exhibition developed by Chicago's Museum of Science and Industry in the mid-1980s, was a critical first step in bringing issues related to women and science into the public eye. As the importance of adults as gatekeepers to girls' science learning emerged, the research in teacher education, parent support, and adult learning became more relevant, as did feminist learning theory. However, though this literature was helpful, it was only after sociocultural ideas about learning and meaning-making became more developed, and the negotiation of identities was articulated as learning, that there seemed to be a coherent link from this work to the arena of family learning in museums.

Fortunately, what has emerged in more recently published literature within the field is the growing awareness of identity and learning as being interconnected and the critical role that a community of practice or sociocultural infrastructure plays in identity development. By using a sociocultural framework to inform our work at The Franklin Institute, we have been able to move beyond the individual as the focus of analysis to the interconnection of individual(s) and community, in ways that take into account shared perspectives, understandings, and co-evolutions. To provide a specific example, our work with *Parent Partners in School Science* (PPSS), discussed earlier in this chapter, has benefited from cutting-edge theories on family engagement in schools proposed in a study by Barton, Drake, Perez, St. Louis, and George (2004). This study has led us to question the existing paradigms about parent involvement in schools and to develop new questions about what parent involvement really means. As we continue to explore the impact of a program designed to build connections between home and school, we have begun to look more closely at what Barton et al. propose—that is, a shift away from the focus solely on *what* parents do to engage with their children's school activities to a focus on *how* and *why* parents are engaged, and the complex *ways* in which their engagement occurs. In our efforts to understand the successes and challenges within PPSS, we are using this parent engagement framework to document the ways in which the program has brought teachers and parents together in support of children's science learning in and outside of school. The consideration of impact within this theoretical framework was compelling to us because it helped us see parent engagement as a dynamic, relational phenomenon, not just about trying to get parents involved in schools, but rather, getting parents and teachers to engage with one another across home and school contexts. The success of this model in facilitating our work suggests that such changes in understanding dynamics and relationships have great potential for influencing the work of the entire museum community.

ENDNOTES

An earlier version of this chapter was published in *Science Education*, 88 (suppl.1): S34–S47, 2004, © Wiley Periodicals, Inc. Used with permission of John Wiley & Sons, Inc.

3

Students, Teachers, and Museums: Toward an Intertwined Learning Circle

Janette Griffin

Effective learning in museums by school students is dependent on the behaviors, attitudes, expectations, and provisions of three main contributors: students, teachers, and museums, including exhibitions, programs, and museum educators.[1] While it is recognized that parents, school curricula, and wider school and museum attitudes toward field trips have impact, this chapter explores the complex and interdependent relationships among students, teachers, and museums, and shows that each forms an intertwined strand of a learning circle. Following a brief overview of our broader understanding of learning in museums, each segment of the learning circle is discussed, with a view to finding an intertwined circle of learning for students. Examples and study findings come from a wide range of museum types from many countries and involve students of all ages and levels of school.

HOW WE UNDERSTAND LEARNING IN MUSEUMS

Visitors to museums value their ability to choose how and what they attend to and exploit this strategy in order to pursue their personal agenda. Learning in a museum involves aesthetic appreciation, growth of motivation and interest, understanding ideas, and social interaction (Schauble, Beane, Coates, Martin, & Sterling, 1996, p. 24). These aspects of agenda, choice, and sharing are consistent with the view of Paris et al. (1998): that museums are environments where people "construct personal meaning, have genuine choices, encounter challenging tasks, take control over their own learning, collaborate with others, and feel positive about their efforts" (p. 271). Falk and Dierking (2000) describe four contexts and the factors within these contexts that are fundamental for museum learning:

- Within the personal context: motivation and expectations; prior knowledge, interests, and beliefs; choice and control;
- In the sociocultural context: within-group social mediation; facilitated mediation by others; and cultural background and upbringing;
- In the physical context: advance organizers; orientation; architecture and large-scale environment; design; reinforcing events; and experiences outside the museum;
- The fourth and essential context of time is overlaid across the above three.

The first three contexts can be linked to the three segments in the learning circle being discussed in this chapter: the personal context of individual students; the sociocultural contexts involving students, teachers, and museum educators; and the physical context of the museum exhibitions and programs. The learning of individual students results from the intermingling of these contexts. This chapter focuses primarily on the influence of the sociocultural context of students, teachers, and museum educators, but inevitably also includes aspects of the personal and physical contexts.

THE STUDENT STRAND

Historically we have tended to look at groups of students on a field trip as a single entity. More recently, there has been increased emphasis on looking at individual students' learning processes and how they can be facilitated, particularly by paying attention to the students' own views of their learning and the environments that encourage learning. Young people have clear views about learning, including the roles of personal interests and motivation, choice, social interactions, and learning styles. They also have strong views about existing learning provisions and opportunities afforded in museums.

Students Are Learning: Agendas, Values, Emotive, and Social Environments

The observation and recording of students' behaviors in exhibitions in Australia have produced surprising results. The tapes revealed that when moving freely the students engage in learning-related conversations for over 80% of their time, with much of their "learning talk" occurring as they move between exhibits (Griffin, Meehan, & Jay, 2003). They link what they see to prior experiences and discuss similarities and differences between exhibits. While a general look at a group of students in a museum gives the impression of constant movement, observation of individual students reveals that they spend considerable time engaged with exhibits, actively involved and sharing learning.

Learning agendas influence individuals' learning as shown by Anderson (1999b) at an interactive science center. Personal attitudes and prior experiences beyond both the museum and school activities influence very different learning outcomes for different members of the same class, and can be the greatest influence on the learning that happens on the visit (Storksdieck, 2006). Prior experiences in the formal learning environment can affect students' perceptions of informal learning environments as places they would enjoy visiting (Taylor, 1996). Jensen (1995) found that when given a choice, children indicate they prefer visiting museums with their families rather than with their school class. In a Sydney study, teachers and students could clearly describe distinct circumstances in which students were learning or enjoying themselves with there being a strong feeling among students and teachers that learning happened in school and in particular involved reading and writing. Elementary students declared that learning and enjoyment went hand in hand when it was fun—they had choice and they were with friends or family. Older children added emphasis on the learning being useful, and actively being involved (Griffin, 1994). These findings agree with the view of Paris et al. (1998) that, "Interest in a topic involves both feeling-related characteristics, such as enjoyment and involvement, and value-related characteristics, such as attributing significance to an activity. . . . When students attribute positive values and feelings to tasks, they are likely to choose them and pursue them vigorously" (p. 280).

While they recognize that a visit to the museum is designed by their teachers to assist their learning, students also want it to be a satisfying social occasion when they learn with and from their peers. When young children are encouraged to interact with their peers, this interaction can be a rich source of motivation for further investigation and learning (Watson, Aubusson, Steel, & Griffin, 2002). Students also want to be emotionally engaged but not emotionally confronted (Ballantyne & Packer, 2002; Groundwater-Smith & Kelly, 2003).

The provision of choice and control in their learning in museums, as well as appropriate opportunities for orientation and rest, are valued by students (Falk & Dierking, 2000; Griffin & Symington, 1997; Paris, 1997; Shelnut, 2000). They need to know why they are going on the field trip and how the information they are gathering will be used. They need some freedom to choose specific aspects of their learning, some ownership of the way in which they are learning, and encouragement to share their learning with classmates and elders, albeit within the framework set by the teacher and/or museum (Griffin, 1998).

Hindering Learning

Bamberger and Tal (2005) found that students were critical of situations where the museum itself offered and encouraged choice but the museum guides prevented it. "[T]he students were passive during the guides' talks that present and discuss unfamiliar concepts" (p. 12) or if the talk did not match the age of the audience or address students' prior knowledge. This frustration was greater in an environment that encouraged choice than in one where this was not so explicit.

In interviews with Year 10 (i.e., tenth grade) students following a trip to a museum, there was discussion about the use of worksheets; two students' views provide insights into the reasons that they did not like them: "because you have to go around looking for the information, you haven't got time to study the things we want to see." "[We didn't learn anything because] most of it went in through the eyes and out through the pen" (Griffin, 1998). Similar aged students, asked what they least looked forward to about an environmental studies trip, answered "boring" talks, worksheets, and assignments (Ballantyne, Fien, & Packer, 2001). The authors noted "that although four of the five programs [being studied] gave students activity sheets to complete during the visit, only 3 out of . . . 440 students mentioned this as something that had contributed to changing their understanding or feelings about the environment" (p. 35).

Students given worksheets behave differently from those who have not. Worksheets tend to narrow the focus of the viewing to only finding the answers in order to complete their assignment rather than using their own curiosity to explore the exhibit. Collecting answers directly from the displays involves mechanical activity and does not encourage connections and analysis (Randol, 2004).

In summary, students' prior experiences and valuing of museums have an impact on their entering agendas. They prefer to work with their friends, to choose the aspects they are looking at in museums, and to exploit their curiosity and interest. They do not like guided tours or worksheets that narrow their opportunities to explore. They like to learn in their own style and in a social environment.

THE TEACHER STRAND

There is considerable discussion in the literature about the role of teachers and the learning opportunities they plan and implement for students on field trips. These issues center on reasons for taking students on field trips and the value placed on them; preparedness to create the conditions that encourage learning during a field trip; relationships between school and museum learning; and teachers' participation in the learning process during the visit.

Why Take Students?

Prior experiences and confidence greatly influence decisions about taking field trips. Teachers choose field trips they think will be educational, that are appropriate to the age and level of their students, and that will enhance curriculum subjects. Being entertaining for the students is also important. Teachers' valuing of the trip has an impact on the learning and what they see as their role in accomplishing these goals (Ballantyne & Packer, 2005; Tal, Bamberger, & Morag, 2005; Tran, 2003).

An interesting investigation in Canberra, Australia, asked teachers to place their preferences for the visit on a grid with two axes (Zerafa, 2000): one axis with *Informing* (focus on content) versus *Inspiring* (information presented creatively, able to explore the subject); the other axis was *Active* (thinking, participating and exploring) versus *Passive* (information placed in front of students, rote learning). Most teachers' preference was the *Active/Informing* quadrant, with a second group choosing *Informing/Passive.* An interesting conundrum appeared in this study's results: "Museums were thought to have high academic relevance for school excursions, but there was also an expectation that they were not a destination that students enjoyed, being 'dull' and 'boring'" (p. 56).

Encouraging Learning

There have been several programs developed that link museum and school learning. An approach developed to help teachers to incorporate museum visits into their classroom-based topics named SMILES (School-Museum Integrated Learning Experiences in Science) was developed and tested in Sydney (Griffin, 1998). This encompassed close integration between school and museum learning; student planning of the aspects of their visit; time to research specific aspects of the class topic; working in small, semi-autonomous groups with freedom to move (within the parameters set by the teacher or museum); the opportunity for physical and mental rests as needed; and a choice of activities that complement the free-choice nature of the learning setting. The factors that seemed to best support learning were purpose, choice, and ownership. The students need to have some purpose for the visit, and they need to know why they are there and gathering the information and how they will use it; they need choice in specific aspects of their learning; and they need ownership over the way they are learning. They also need to be able to share their learning with classmates and elders in a comfortable social environment. Orion (1993, 1999) developed a similar model for integrating field trips into a curriculum unit, with the field trip occurring relatively early in the unit.

Xanthoudaki (1998) describes the contrast between art museum programs that are aimed at supporting the teacher by "helping teachers help themselves" and ready-prepared programs that may or may not match the requirements of teachers or students. She describes these two models as "the gallery as classroom resource" (Model 1) and "the gallery as teacher about its own collections" (Model 2). She found that visits to a gallery using Model 1 were more likely than the second model to lead to incorporation of the visit into classroom art "because of the correspondence between educational programs in the institution and school curriculum requirements" (p. 189).

Hindering Learning

Unfortunately, the reality of many field trips involves teachers who visit with few and often poorly defined objectives, little preparation, and loose connections to the classroom unit or curriculum, primarily because the setting is chosen first rather than the learning topic (Storksdieck, 2006). Gammon (2001) lists barriers to student learning in a museum: activities where there is no obvious reward or motivation for continuing; activities poorly matched to the abilities of the audience; activities that make visitors look foolish; and activities that preclude social interaction. Teachers' agendas and management of these issues can also greatly influence students' experiences on field trips.

Teachers are rarely or only superficially provided with useful professional preparation for taking field trips during their initial training or during their teaching career. There is an assumption that everyone knows what to do. In fact, what generally happens is that teachers behave in the same way as their teachers did when they were at school, and this is the only model current teachers have available to them. Additionally, taking students on field trips is very stressful due to the logistics of arranging the visit, the transportation, and, if needed, substitute teachers back at school, gaining permission from school administrator and parents, concern about safety and student behavior in an environment that does not have four walls. It is not surprising that attention to learning processes gets left behind or that teachers "hand over" responsibility for the visit to museum staff, particularly if they are uncertain themselves about the content in the museum.

Despite some uncertainty about specific content, teachers generally have better knowledge, higher awareness, and greater openness about museums per se than their students. Hence, teachers experience the visit to an out-of-school learning environment quite differently from their students and, in fact, can create a barrier to student learning by only selectively and often insufficiently passing on information about the site (Storksdieck, 2001).

Robins and Woollard (2003) found that art and design teachers in London "often perceived a distinct change in their role when they took students to museums and galleries . . . characterised as moving away from the teaching activities of the classroom (demonstrative, informative, activity-based) toward more passive and organisational roles (supervisory, observational, pastoral)" (p. 4). They considered the environment to be beyond their control, and were concerned that museum staff or students might discover their lack of understanding or knowledge. They felt self-conscious about talking in public spaces, and found it difficult to assume a teaching role in this context; thus, they often attempted to apply a classroom practice template because they did not have the experience and confidence to employ alternative approaches.

One of the most commonly cited causes for hindering learning in museums is the use of worksheets, often used as a "controller" of the students and their learning. They can be useful if well designed and well used. Teachers tend to have one of two clear agendas with regard to worksheet use in museums: a survey agenda and a concept agenda (Kisiel, 2001). Those with survey agendas (the majority) use worksheets that have little to do with the classroom curriculum. This dampens motivation, reduces opportunities for linking experiences with prior knowledge, and reduces control and choice in learning. Those with concept agendas, on the other hand, use broader questions and clearly link the visit to the concepts being studied at school.

Parsons and Muhs (1994) showed that too often "worksheets interfere with the social interaction and maybe learning. . . . We found that while filling in worksheets, group members spoke less to one another, looked at the exhibits less, usually gave up on the worksheets during their tour and didn't spend any more time in the aquarium than groups without worksheets" (p. 60). Teachers' expectations that students should complete the prepared sheets act to impede student learning. Students declare that they feel rushed as they seek to find the answers to the questions that had been set for them (Groundwater-Smith & Kelly, 2003).

The "hands-off" nature of teacher behavior during field trips is a common phenomenon. No teacher interviewed for a study in Israel was an active facilitator, and in many cases the teachers had no idea of the field trip program or rationale—teachers were minimally involved in planning and conducting the museum visit (Tal et al., 2005). A further concern that unfolded in this study, which is apparent elsewhere, is an increase in subcontractors and bus companies who plan tours and make all the arrangements for museum trips. This trend further distances the teacher from student learning.

Overall the picture is mixed at best. The major issues with regard to teachers' facilitation of learning in museums are the overriding sense of stress over organizational and management issues and the lack of understanding of the appropriate pedagogical approaches. Teachers rarely encourage or provide opportunities that tap into either the personal or the sociocultural learning contexts (Falk & Dierking, 2000) of students, and hence are inhibiting much potential learning. These issues can be solved. The best pathway may be to develop closer collaboration between teachers and museum educators.

THE MUSEUM STRAND

Exhibitions and Visitors

The museum exhibition developer (or team of developers) can control the exhibition and what can be seen and potentially done by interacting with it, but the mental distance between the desired activity and the visitor may be difficult to determine, and it is these connecting steps that influence the outcomes of the visitor (Ansbacher, 2002a). If the experience needs to be interpreted to become meaningful, and the interpretive steps are not apparent, then the exhibition becomes superficial at best and meaningless at worst. Students are rarely considered in this mix.

Exhibitions "require integrated settings that foster discussion, challenge the learner, make connections to issues of interest to the learner, and provide guidance for applications in the world outside the museum" (Hein, 2004, p. 422). Anderson (2004) describes features in major museums in London that he considers support learning: engaging the audience in debate through use of poetry, questions, and gallery books; use of computers and other interactives; community programs; audio points in exhibits; information rooms; curator signature on wall text; study and discovery areas; and interaction with exhibitions. Inclusion of these features in the British Galleries of the Victoria and Albert Museum doubled the number of visitors who said they learned something, and the average stay time in the galleries increased from 8 to 54 minutes. Rand (2001) formulated a Museum Visitors' Bill of Rights, which includes the rights to comfort, orientation, welcome, enjoyment, socializing, respect, communication, learning, choice and control, challenge and confidence, and revitalization.

Students are rarely given the opportunity to share their perspectives about exhibitions; however, Groundwater-Smith and Kelly (2003) used an innovative methodology to investigate students' perspectives on a museum in Sydney. In teams, students took photos of elements that assisted or inhibited their learning, then made conceptual posters using these images. Cognitive and physical factors were raised: students wanted to know how things work; to think through ideas; to have opportunities to ask questions; to be able to handle, manipulate, and closely examine artifacts and exhibits; to use a variety of sources of information in language that is appropriate to their age; and to be stimulated through various senses. They also need to feel safe and comfortable, to move around readily, unimpeded by prohibitive signs, in well lit and inviting spaces. They welcomed opportunities to be "fully engaged with provocative questions, fascinating and puzzling exhibits and clear, well structured and accessible information." (p. 19)

Sacco (1999) involved teachers in evaluating different exhibit-related experiences in terms of their usefulness as a teaching tool in a museum setting. Teachers liked multisensory exploration of museum objects and live demonstrations. They were not so positive about computer and video presentations or guided tours.

Museum Educators, Exhibitions, and Student Programs

There has been an assumption that school students in museums passively receive the given wisdom presented through the exhibitions or by museum staff. There has been a long tradition of exhibitions being developed by curators and designers then "handed over" to the museum educators to "interpret" these exhibitions for the audience. Fortunately, there are increasing numbers of museums that develop exhibitions and programs using a team of experts from many fields. A less common practice is sharing expertise through genuine collaboration between teachers and museum educators to enhance learning for students. The responsibility for providing successful museum-learning opportunities falls on the shoulders of museum educators. However, museum educators largely learn on the job; there are few specific training courses available. As a result, many programs are "teacher centered," with minimal student interaction (Tran, 2003).

Mathewson and McKeon (2002) suggest that there is a misrecognition of the power relationships between museum educators and teachers, particularly in art museums, that "enables the reproduction of relations in which museums are dominant and in which school-based educators have an ill-defined and often educationally ineffective pedagogical role" (pp. 1–2). De Witt and Osbourne's *Framework for Museum Practice* (in press) is based on an activity theory that predicts that museum educators and teachers are operating in different activity systems or different contexts. Robins and Woollard (2003) found that museum educators considered a successful visit to be one where the students find personal relevance in the exhibitions and feel comfortable in the gallery, while teachers feel a successful experience has direct influence on students' practical course work. Museum educators were critical of teachers' approaches, particularly their lack of involvement and lack of knowledge. "The research does suggest the need for greater understanding on both sides of the educational context in which the other works, and strategies to assist collaboration at a deeper level to provide for the learning needs of pupils" (Robin & Woollard, 2003, p. 12).

While there is considerable emphasis placed on electronic or printed material supplied by museums for teachers, little study has been conducted on these and this is an area requiring more work. Most written or electronic material developed by museum educators for teachers tends to focus on the organizational aspects of a visit and very little pedagogical advice is given (Storksdieck, 2006).

In an attempt to close the gap between museum educators and teachers, Guy and Kelley-Lowe (2001) report on a program where preservice teachers worked with educators in a science center and gained a wide range of insights into teaching science and the value of science centers as learning environments. An excellent partnership program of teacher ambassadors is described by McLeod and Kilpatrick (2002). Ambassadors are selected from their school to be the person who forms a link with local museums: "When the centers and the school districts work together to develop inquiry based learning opportunities linked to the school curriculum, the window of opportunity for making students' learning more meaningful, more connected and therefore more permanent, opens wider" (p. 62).

The issue that stands out in regard to relationships between museum educators and teachers is that the pedagogy is uncertain on both sides. There is considerable room for more work both in research and in professional development. Schools and museums have different but complementary roles, and there is a need for closer alliances between these institutions and their staff.

TOWARD AN INTERTWINED LEARNING CIRCLE

This chapter has explored the three strands that form an intertwined learning circle allowing students of all ages to enjoy learning in museums. A number of ideas have been provided for better collaboration that may lead to effective learning opportunities. The development of a range of museum-learning pedagogies and provisions to match different settings, circumstances, and learners is clearly needed. This involves the physical context—recognition of the unique qualities of the museum itself, its exhibitions, and programs; the personal context—teachers' and students' agendas, values, emotions, prior experiences, knowledge, and expectations; and the sociocultural context within which the three strands meet: students, teachers, and museum educators. It is clear that the way forward is to weave the three strands into a full circle through understanding, sharing, mutual respect, and collaboration. This involves an exciting challenge: listening to the needs, wants, and concerns of each group, learning from each other, and working collaboratively toward a shared goal of enjoyable and meaningful learning experiences.

ENDNOTES

An earlier version of this chapter was published in *Science Education*, 88 (suppl.1): S34–S47, 2004, © Wiley Periodicals, Inc. Used with permission of John Wiley & Sons, Inc.

1. "Museum educators" is used to refer to paid and voluntary staff working with students.

4

Exhibit Design in Science Museums: Dealing With a Constructivist Dilemma

Sue Allen

A LEARNING DILEMMA

At first glance, the exhibition space of a science museum seems an appealing educational alternative to a school science classroom: hands-on exhibits are novel, stimulating, evidence-rich, multisensory, and fun. The environment provides myriad personal choices, without any teachers enforcing agendas, without curricular constraints, without testing or accountability.

But science museums are actually very difficult environments to engineer for learning, precisely because of these same attributes. In school, a teacher can regulate her students' progress through challenging material, ensuring that they all arrive at the rewarding or significant climax of a lesson. By contrast, if an exhibit has a boring or confusing component, visitors have no way of knowing[1] whether persisting will be worth the effort; and in an environment full of interesting alternatives, they are likely to just leave the exhibit and move on. It is not enough for an exhibit to have a culminating point that is rewarding; *every intermediate step* in the visitors' experience must motivate them to continue investing time and attention there. This reliance on continuous personal motivation makes it incredibly difficult to get visitors to follow a narrowly constrained learning agenda, particularly one involving sequential steps or high levels of intellectual effort.

Such challenges form part of a constructivist dilemma: We expect museums to provide a hugely diverse visiting public with the freedom to choose their own paths and create their own meanings. Yet we also want museums to be respected educational institutions where people can spend an hour and come away having learned some aspect of science as currently accepted and practiced. This dilemma plays out at every grain size, from the largest organizational tensions between market and mission to the smallest design challenges of a single-exhibit element.

The past decade of visitor studies has shown that it is, in fact, possible to create exhibit environments where visitors are simultaneously in a constant state of free choice and in the process of learning at least some aspect of science. Some examples from the field have recently been compiled by McLean and McEver (2004). But the work is difficult, and calls for a program of research that focuses on the detailed features of the physical environment in which such learning is deeply situated. In the following sections, I summarize several key considerations for designing such environments effectively. I draw many examples from the past decade of research and evaluation studies at the Exploratorium, where I have most personal experience.

"IMMEDIATE APPREHENDABILITY": CREATING EFFORTLESS BACKDROPS FOR CHALLENGE

Cognitive and sensory overload is a huge problem in museums of all kinds (e.g., Maxwell & Evans, 2002), but perhaps especially in hands-on science museums. Consider the challenge: visitors face a gyrating landscape of hundreds of exhibits, few of which they have seen before, and none of which have standardized controls, mechanisms, or explanations. Adults wanting to support young children must look at each novel device, decipher instructions, guide their children's physical activity, make sense of the experience themselves, translate the significance of it for their children, assess the result, and make adjustments to optimize their children's learning. Over and over, every few minutes, adults coach their children in a range of cognitive and affective skills, with no previous on-site training. It is no wonder that visitors can only engage deeply with exhibits for a limited period (typically about 30 minutes) before they lose focused attention and begin to "cruise" (Falk, Koran, Dierking, & Dreblow, 1985). Museum fatigue critically limits the degree to which visitors can effectively learn any form of science.

A powerful and general approach to reducing the degree of cognitive overload is to design for "immediate apprehendability." By this, I mean the property of an object or environment such that most people introduced to it for the first time will understand its purpose, scope, and properties almost immediately and without conscious effort.

User-Centered Design

At the object level, user-centered design (also called "end-user" or "natural" design) promotes the creation of objects that, by virtue of their physical forms and locations, invite certain kinds of use and not others. Such design often goes unseen and unappreciated because, ironically, masterful design results in objects that seem obvious and simple to use. Started by industrial and human factors designers (e.g., Dreyfus, 1967; Papanek, 1971), user-centered design came into mainstream psychology through the work of cognitive scientist Donald Norman. He used the term "affordances" to refer to the directly perceivable properties of objects that determine how they could possibly be used. "Knobs are for turning. Slots are for inserting things into. . . . When affordances are taken advantage of, the user knows what to do just by looking: no picture, label, or instruction is required. Complex things may require explanation, but simple things should not" (Norman, 1988).

Beyond the concept of affordances, other well-tested principles from this field have been found relevant in museums:

- Using natural mappings to take advantage of physical analogies and cultural standards. For example, turning a knob clockwise increases the amount of whatever is being adjusted.
- Limiting the available controls to one small set at a time. While this can be done most easily with electronic interfaces, it is also possible with physical exhibits. One way is by making controls differentially salient, making the primary ones large or centrally located and the secondary ones smaller or positioned off to the side for later discovery.

- Giving clear feedback in response to manipulation. When an aspect of an exhibit has been manipulated, there should be some form of immediate feedback so that visitors know they have effected a change.
- Standardizing for consistency to reduce cognitive load for subsequent interactions. At the Boston Museum of Science, for example, the Hearphones (small earpieces that provide audio versions of label text) are always positioned in the same place on the exhibit label so that blind or low-vision visitors don't have to search for them each time.

Familiar Activities

On the larger scale, immediate apprehendability may be achieved by designs that remind visitors of familiar activities. Within a few seconds, visitors understand the context of the situation and can attend to interpretation of the specific rules and objects involved. For example:

- Making a complex machine work. *Bike Cycle* is a robotic simulation of a bicycle rider, with hydraulic pistons in place of major leg muscles. To make the robot pedal, visitors must press four buttons in the right sequence and with exact timing, a significant challenge usually requiring multiple attempts. Yet evaluation showed that 70% of visitors said they thought the level of challenge was about right. The exhibit offered visitors immediate apprehendability at the large scale (the robot is immediately recognizable as a simulated person on a bike), as well as the small scale (the controls are four large buttons, centrally located, easy to push, and color-coded to match the relevant pistons).
- A competition, especially a race. *Downhill Race* is an exhibit in which visitors race wheels of different shape and weight down an inclined plane. The presence of two lanes side by side, the exhibit title, and the set of six disks with slightly varying features, all contribute to visitors' immediate apprehension that the exhibit involves racing the disks down the plane. This frees up visitors' time and attention from figuring out "what this thing does," and allows them to explore the scientific principles at the heart of the exhibit—identifying characteristics of the disks that affect their speed. Also, because visitors recognize the activity, adults can immediately support their children

in learning relevant scientific procedures, such as setting up a "fair" race and judging its outcome. A version of this exhibit was deeply analyzed by Rowe (2002), who noted the power of a competition or race as an idea that structured visitor activity.

- Watching and waiting. A familiar activity need not be energetic; high levels of learning were elicited by an "empty" tank of frogs in one of the Exploratorium's temporary exhibitions. The tank contained a mini-ecosystem of pebbles, water, and plants, and was in fact full of frogs, but being nocturnal, most were hidden and sleeping during the museum's open hours. Surprisingly, a study of visitors' conversations revealed that this exhibit was the most likely of any in the exhibition to elicit complex inferences from visitors, such as theories about camouflage and froggy lifestyle (Allen, 2002). It was also the second most likely element to evoke stories, personal associations, and previous knowledge about frogs. Apparently visitors recognized the exhibit as something where one had to wait and watch in order to be rewarded, and they accepted the challenge and used it as an opportunity to share stories and knowledge.

Familiar activities can also be used at the scale of an entire exhibition. For example, the detective puzzle used to frame the *Whodunit* exhibition about forensic science allowed visitors to immediately apprehend the purpose and relationship of distinct elements such as the crime scene, DNA lab, and simulated autopsy laboratory (Walter, 2004).

Immediate Apprehendability on All Scales

Csikszentmihalyi and Hermanson (1995) characterize ideal learning at exhibits as initially driven by curiosity and interest, and then sustained via a "flow" state, in which visitors become fully involved with mind and body in an intrinsically motivated activity. To create a flow state, activities generally need a level of challenge that closely matches the person's skills, as well as a clear set of goals and rules. However, I believe we need to optimize not just the level of challenge but also its timing and context; research suggests that visitors will only engage in a challenge if they are comfortable and oriented (e.g., Hayward & Brydon-Miller, 1984). Thus immediate apprehendability may be particularly important during the early stages of an experience because it lessens distractions and provides visitors with a comfortable framework from which to be curious.

In fact, one might view the whole museum visit as an alternating series of challenges and immediate apprehensions on different scales: A family on holiday chooses to visit the Exploratorium because they are curious to see something novel and interesting. They drive to the museum, expecting its location to be immediately apprehendable, but due to poor road signs they spend time and energy finding the building and even the entrance. Once inside, the admissions desk and bathrooms are immediately apprehendable, allowing them to move effortlessly into the main exhibition space. They notice the movement of a nearby exhibit, a large compound pendulum with arms that flail chaotically. Fortunately the control for the exhibit, a large brass knob, is immediately apprehendable and they turn it to see what different kinds of behaviors they can provoke from the exhibit. Now curious about the point of this strange object, they scan the label and easily identify "What's going on?" as the relevant part to read. They invest time and effort in reading and making sense of this scientific explanation.

Normally, neither staff nor visitors would tell this story by including the immediately apprehendable aspects of the environment. By definition, they are the things we don't notice. But they are just as important from the design perspective because they reduce the ever-present cognitive load on visitors, freeing them to focus on aspects of the environment that are worthy of their attention. Effective orientation to the mundane aspects of museum facilities has been directly linked to higher levels of science learning by children on school field trips (Balling, Falk, & Aronson, 1992). And more recently, Maxwell and Evans (2002) link the physical environment to learning through processes such as cognitive fatigue and anxiety, and they provide evidence that learning is enhanced in quieter, smaller, better-differentiated spaces.

Immediate apprehendability, with its emphasis on comfortable and effortless understanding by users, might seem antithetical to a museum whose learning model rests on the motivating properties of surprise, cognitive dissonance, and curiosity. However, based on studies at our museum and many others, I believe that it forms the essential backdrop that makes such exhilarating foreground experiences possible.

PHYSICAL INTERACTIVITY: A MORE CRITICAL LOOK

Physical interactivity, the ability of an exhibit to respond to visitor actions, is considered a "cardinal feature" of science (and children's) museums. Research on visitor learning in museums suggests that interactivity can promote engagement, understanding, and recall of exhibits. For example, Maxwell and Evans (2002) cite evidence that both children and adults recall actions they themselves perform better than those they observe. Richards and Menninger (1993) found that specially designed interactive galleries at the J. Paul Getty Museum had greater holding times for visitors. Borun and Dritsas (1997) identified interactive design features such as multioutcome and multimodal as key ingredients in exhibits that foster family learning at science centers.

However, "more is not necessarily better" when it comes to interactive features. For example, Allen and Feinstein (2006) varied the level of interactivity in an exhibit displaying microscopic worms. They found that visitors learned more when interacting with live worms than when watching videos, but adding higher levels of interactivity (changeable lighting, focus, and dish location) did not further enhance the experience. Allen and Gutwill (2004) identified five common pitfalls of designing exhibits with multiple interactive features, suggesting that exhibits may have an optimal degree of interactivity and that formative evaluation is essential for ensuring that the interactive features work together harmoniously. Some of the Exploratorium's most attractive and sustaining exhibitions have used little or no physical interactivity at all (for example, *Energy from Death*, displaying animals in various stages of decomposition; or a display of seven optical illusions in the *Seeing* collection). In a temporary exhibition on *Frogs*, the exhibits of live animals were not only more attractive than the physically interactive exhibits but also evoked more frequent and diverse visitor conversations, showing evidence of learning. The single exhibit that evoked the most diverse learning conversations, and the longest holding time, was neither live nor interactive, but a video (Allen, 2002).

A number of museums have been investigating ways to create interactive exhibits that go beyond "hands-on" manipulation to support visitors conducting their own investigations and making sense of the results (e.g., Bailey, Bronnenkant, Kelley, & Hein, 1998; Sauber, 1994; Schauble & Bartlett,

1997). Most have used layered exhibit designs that combine easy initial access to phenomena with opportunities for deeper cognitive experiences. Most recently, the Exploratorium created exhibits that supported "active prolonged engagement" (APE) by visitors: long holding times, varied intellectual engagement, and greater use of the exhibit (rather than the label) to answer visitors' own questions. Exhibits that achieved these goals fell into four broad categories: those that invited visitors to explore a phenomenon, to figure something out, to observe a beautiful phenomenon, and to construct things from smaller elements. Particularly successful strategies for exhibit design included breaking exhibits into multiple stations to support group engagement, and posing challenges in exhibit labels (Humphrey, Gutwill, & the Exploratorium APE Team, 2005).

Overall, several studies suggest that, while recognizing the power of interactive experiences, we should be skeptical about sweeping claims that interactivity is essential to learning, or even that it necessarily creates the most powerful, memorable, or attractive experiences in our museums. In addition, it has become clear that interactivity is a multifaceted notion: exhibits can offer far more than opportunities for physical manipulation, but designing for deeper engagement and self-authored learning by inquiry is extremely complex and requires nuanced design and extensive evaluation.

THEMATIC COHERENCE: AN ONGOING CHALLENGE

Most science museums arrange their exhibits to support thematic coherence—that is, hoping that visitors will perceive the intended common ideas or themes. However, visitors tend to be highly literal in their interpretations of exhibits (see, for example, Gammon, 1999), and do not easily infer abstract concepts from multiple individually designed exhibits (see, for example, Klages & Exploratorium Staff, 1999).

In the past decade, museums have continued to explore various techniques for making abstract concepts and themes more apparent to visitors, including linear sequencing of exhibits, unified design aesthetics among exhibits in a cluster, advance organizers (conceptual or spatial overviews) at the entry to thematic collections, and labels that echo and reinforce the abstract theme—not to mention additional mediational tools such as audio tours and the presence of docents. None of these are new, but their effectiveness depends on their precise design, and their use in combination brings in new, untested possibilities for visitor learning.

The Exploratorium is one of the institutions that has conducted numerous experiments with the coherence of exhibit collections. The *Seeing* exhibit collection, which partly replaced an earlier collection on *Light, Vision, Color, and Optics*, was found to be successful in shifting the theme perceived by visitors toward the idea of seeing as personal interpretation, dependent on attention (Whitney & Associates, 2003). On the other hand, the *Traits of Life* collection was relatively unsuccessful in communicating its abstract theme (viz., four characteristics common to all living things). Most visitors interpreted the display of a range of living creatures as a celebration of diversity rather than commonality (Hein, 2003), and many misidentified the intended themes of the collection as "cycles of life" or "environmental interdependence."

Visitors may simply not expect to look for strong interexhibit links; studies of several exhibitions have shown that explicit connections among exhibits are relatively rare in visitors' conversations (Allen, 2002; Leinhardt & Knutson, 2004). Achieving thematic clarity for exhibitions may be particularly difficult in environments where most exhibits are, by design, loosely related via a web of alternative possible connections. Open spatial layouts may even undermine thematic coherence to some degree: Allen (2003) showed that simply adding partitions around a group of related exhibits increased the percentage of visitors who identified their intended common theme from 30% to 51%.

Thematic coherence, especially of abstract concepts, is often difficult to achieve on the open floor of science museums. Abstract synthesis is highly effortful, and the rewards for engaging in it may not be obvious. "Humans prefer to add new experiences, new data and new information to conceptual schemes they already have, rather than to accept new overarching ideas" (Hein, 2003, p. 36). Yet, because of its importance, this aspect of informal science learning seems worthy of more experimentation. Driver and colleagues, when defining constructivism in classrooms, argued, "Learners need to be given access not only to physical experiences but also to the concepts and models of conventional science" (1994, p. 7). Perhaps educational research in other settings could have some relevance for museums here. For example, it is well known that learners may need explicit modeling and coaching when learning new skills, yet exhibitions rarely provide visitors with guided opportunities to practice making and discussing thematic connections among the exhibits they visit.

DIVERSITY OF LEARNERS: HELPING VISITORS OPTIMIZE THEIR OWN LEARNING

If learning is fundamentally visitor driven (the basic constructivist position), how can a static, unmediated exhibit collection possibly support the huge diversity of learners who visit a science museum?

Over the past two decades, several theoretical approaches addressed learner diversity directly. The theory of multiple intelligences (Gardner, 1993b) proposed that there are different cognitive styles for understanding the world, not limited to the verbal and logical-mathematical. A second was the "learning styles" systems proposed by Kolb (1984), McCarthy (1987), and others, which classified learners according to their preferred modes of perception and processing of information. A third was the recognition of different sensory modes of experience: visual, auditory, tactile, smell, and even taste. These approaches validated what many museum educators already knew: visitors vary in their preferences, styles, and motivations for learning.

Museums have attempted to design a broader range of learning experiences in response to this diversity. In one study, Borun & Dritsas (1997) determined that one of the seven characteristics of exhibits that facilitated learning by family groups was "multimodal," meaning that they appealed to different learning styles and levels of knowledge. Another key study validated the effectiveness of universal design, whereby objects and environments are designed to be usable by people with the broadest possible range of abilities. Specifically, staff at the Boston Museum of Science modified a hall of dioramas, adding objects that afforded touching, listening, and smelling, and activity stations that invited visitors to make comparisons between animal body parts and human tools. These changes improved the experience for all visitors: there were increases in holding times, label use, and understanding of the area's major themes (Davidson, Heald, & Hein, 1991). A recent survey by Tokar (2003) suggests that designing for physical access has become common among institutions with hands-on science exhibits. However, universal design with its larger goals, including intellectual access and participation for all visitors, has not yet been integrated into mainstream design.

Another technique the field has explored to engage a more diverse audience is the use of narrative in exhibits. Narrative, loosely defined as personal storytelling, has long been recognized as a powerful exhibit technique in historical and cultural museums (Bedford, 2001; Rounds, 2002) and is even regarded as a fundamental mode of human thought by some psychologists (e.g., Bruner, 1986). In an influential argument for constructivism, researchers such as Roberts (1997) and Silverman (1995) have proposed that narrative, particularly with multiple voices, should replace authoritative knowledge dissemination as the iconic mode for museums to accomplish their educational mission. However, narratives have been much less prominent in science museums, where the dominant mode is still hands-on inquiry with a single authoritative explanation. Martin and Toon (2005) suggest that narrative may be particularly effective in helping to reduce the alienation that many people feel from "the priesthood" of science.

Evaluations at the Exploratorium have suggested that visitors readily engage in telling or listening to stories on emotional topics such as love, AIDS, or the memories associated with precious objects from one's home country (McLean, 2003; Pearce, 2003). Far more difficult is creating compelling stories that make phenomenon-based science exhibits more personally meaningful to visitors, as was tried in the *Finding Significance* study (Allen, 2004). Part of the challenge was that visitors wanted narratives to be tightly connected to their immediate experience of each exhibit; they were less engaged by a story about a related situation, even an exciting one. Another obstacle to incorporating narratives into exhibits is that the museum floor is often noisy and chaotic, and not conducive to sitting and listening. Experiments with quieter areas, using selective lighting, and partitioning the spaces have been noticed and appreciated by Exploratorium visitors (Gutwill-Wise, Soler, Allen, Wong, & Rezny, 2000; Hein, 2003).

Finally, an underutilized approach that might also support visitor diversity is helping visitors to assess their own learning. Research in schools has shown that self-assessment and metacognition (the ability to monitor and control one's own thinking) are key learning skills; in the museum context, they might be particularly useful tools for visitors to identify their own styles and preferences for learning, and to be clearer about what they know and where they want to learn more. Visitor researcher Serrell emphasizes "people appreciate being given information that will help them make intelligent choices" (1996, p. 72). Perhaps this could be taken a step further to include activities and experiences for learners to understand their own learning more fully.

CONCLUSION

Designing science museum exhibits that engage visitors moment-by-moment and yet also support their science learning is a difficult but not impossible task. Given the challenges, it seems critical to support the design process with a strong program of research and evaluation, which is an established practice at the Exploratorium and an increasing number of museums worldwide.

Many studies suggest that more attention needs to be paid to user-centered design at all scales, from the precise affordances of exhibit controls to the layout and orientation of the surrounding environment. A useful concept for framing the research on museum learning is that of immediate apprehendability, which reduces cognitive overload and frees visitors to focus on those aspects of the environment that are worthy of their attention. We already know some methods for achieving immediate apprehendability, such as basing exhibits on activities that are already familiar to visitors, but there is much to learn. For instance, we need to know more about the impacts of apprehendability versus challenge at different scales in the museum visit, at different times, and with different audiences.

Research has shown that physical interactivity is not a simple and universal prescription for effective learning. While it often enhances visitor engagement, understanding, and recall, interactivity is not always essential to a powerful and sustaining experience, and too many interactive features may even hinder visitors' learning. A number of institutions have extended the concept of interactivity to include extended engagement and self-directed inquiry, using exhibits that engage multiple users and offer multiple intriguing outcomes. Recent work on forms of gesture and play (e.g., Rennie & McLafferty, 2002; vom Lehn, Heath, & Hindmarsh, 2002) have highlighted the importance of nonverbal learning, but more research needs to be done in this area because this may be the dominant form for a three-dimensional physical interaction, especially for children.

It is often difficult to make thematic connections apparent to visitors (especially of abstract scientific concepts or principles), in part because visitors frequently learn in a literal way at exhibits. Environmental factors, such as the presence of partitions surrounding an exhibit group, can help to make thematic connections clearer, but there remains a vast arena of useful research that could be done. How can we be more effective in manifesting the abstract concepts and principles of science, providing significance for a potentially fragmented set of experiences? What combination of exhibit and environmental design features provides the most support? Spatial and environmental designs are multifaceted and difficult to prototype and test, so much remains unknown in this area.

In 1992, the American Association of Museums published *Excellence and Equity*, a call to museums to recognize education as central to their public service, and to be more inclusive of diverse people in all aspects of their operations and offerings. One part of this charge is to create more inclusive exhibit collections. Research has shown that incorporating the principles of universal design, incorporating a broader range of approaches and sensory modalities, and including alternative approaches such as narrative can contribute to this effort. In addition, several researchers have proposed specific techniques to make prototyping and evaluation more culturally responsive, bringing in those members of the audience who may normally be excluded (e.g., Garibay, 2005; Reich, 2005).

As museum staff, we also need to be explicit about our own intentions in terms of what aspects of our learning environment we regard as *appropriate* for visitors to struggle with. Should we make explanations of scientific phenomena easy to locate and understand, or do we want visitors to rise to the challenge of investigating phenomena on their own terms? Should we create sequenced exhibits and linear paths to reduce the effort of navigation and connection-making among exhibits, or should we keep the floor plan open because connection-making is exactly where we believe visitors should be spending their effort? These are ongoing questions requiring institutional prioritizing as well as further research.

Effective design often takes many iterations, even for simple devices such as telephones and radios. How much more challenging, then, is the design of a unique and novel exhibit that must be robust, easily usable by people of any age and background, and lead to the learning of some aspect of science or the world in a personally meaningful way? In the face of irreducible complexity of both physical systems and humans, we are unlikely to ever create generalizable enough design principles to obviate the need for research, prototyping, and evaluation. Much of this work will require careful and detailed study at many scales if we are to understand the myriad alternative ways in which visitors experience, interpret, and learn from our exhibition spaces.

ENDNOTES

The past decade of exhibition development at the Exploratorium, from which many of the examples in this chapter come, was conducted under the leadership of Kathleen McLean, then director of the Center for Public Exhibition. She also supported and guided the evaluation and research described.

An earlier version of this chapter was published in *Science Education*, 88 (suppl.1): S34–S47, 2004, © Wiley Periodicals, Inc. Used with permission of John Wiley & Sons, Inc.

1. This chapter focuses on the use of exhibits by visitors in "self-guided" mode (i.e., without mediation by staff members), because this is the most common way visitors to science centers experience exhibits.

5

Research on Learning From Museums

Léonie J. Rennie and David J. Johnston

In this chapter, we take stock of research on museums as learning institutions. We explore the recent antecedents to our present perspectives, identify what seem to us to be the key issues relating to learning, and draw implications from these for future progress in the field of museum research. In particular, we emphasize the need for a holistic approach to research, encompassing the personal, contextual, and time-related aspects of a museum experience.

The 1994 conference *Public Institutions for Personal Learning: Establishing a Long Term Research Agenda* (Falk & Dierking, 1995) brought together contemporary perspectives espoused by researchers about the nature of learning and how these perspectives might influence research on learning in the museum field. Already, publications such as those by Falk and Dierking (1992) and Resnick, Levine, and Teasley (1991) had laid the epistemological foundations for the next decade of research. In *The Museum Experience*, Falk and Dierking (1992) introduced the Interactive Experience Model, which emphasized the importance of the context for learning in museums. In their volume, *Perspectives on Socially Shared Cognition*, Resnick et al. (1991) brought together researchers from psychology, sociology, anthropology, education, and linguistics to examine the implications of viewing cognition as a social phenomenon. As Resnick et al. pointed out, "Our daily lives are filled with instances in which we influence each others' constructive processes by providing information, pointing things out to one another, asking questions, and arguing with and elaborating on each other's ideas" (p. 2).

Allied perspectives had been developing in the previous decade or so, often with roots in the ideas of Vygotsky and, less explicitly, Dewey. Lave and Wenger (1991) developed their conception of situated learning through studies of apprenticeship, arguing that newcomers to a field attain membership in a community of practice through "legitimate peripheral participation," their descriptor for "engagement in social practice that entails learning as an integral component" (p. 35). Focusing more directly on formal schooling, Brown, Collins, and Duguid (1989) wrote about situated cognition, collaborative learning, and cognitive apprenticeship. The term "apprenticeship," they claimed, emphasized "the centrality of activity in learning and highlights the inherently context-dependent, situated, and enculturating nature of learning" (p. 39). Around this time, mainstream educators were beginning to recognize that in-school education might benefit from some of the work being done on learning out of school (see, for example, the 1987 American Education Research Association Presidential Address [Resnick, 1987]). In this vein, Brown et al. (1989) noted that "learning, both outside and inside school, advances through collaborative social interaction and the social construction of knowledge" (p. 40). With these emphases on the contextualized or situated nature of learning and socially shared cognition contributing to our understanding, it is not surprising that the largest growth of research in museums in the past decade has been based on sociocultural premises (see, for example, Schauble, Leinhardt, & Martin, 1997). This research is contributing greatly to our understanding of how visitors make meaning of their visit experience, and deepening our understanding of what and how learning happens in museums. There is still much to learn, however, so we must continue to seek more effective ways of investigating the museum experience and how it may affect people's lives.

LEARNING IN THE MUSEUM CONTEXT

If museums and similar institutions are to have an impact on people's lives, then they must change people in some way. A person watching a cuttlefish for the first time at an aquarium might be struck by how rapidly it can change its color. At another exhibit, a visitor may be shocked by the number of marine species threatened as pollution destroys the coastal sea grass, and subsequently join an environmental group. Both people experienced a change. The first, when helping a child with homework at a later time, might simply remember that cuttlefish camouflage themselves. The second may build upon the aquarium experience and become committed to conservation. And many other visitors may experience impacts between these two extremes.

We argue that these changes, or impacts, involve learning. There are many models or theories about learning, and Hein (1998) and Matusov and Rogoff (1995) provide useful discussions in the context of museums. To strengthen our research on learning from museums, we need to identify the essential attributes, or characteristics, of learning that can accommodate most, if not all, of these theoretical perspectives, and then explore their implications for research into the impact of museums on people's lives. To this end, we propose three characteristics of learning: first, learning is a personal process; second, learning is contextualized; and third, learning takes time. Of course, these attributes are not mutually exclusive because together they characterize learning, but we will describe them separately in an endeavor to draw more clearly their implications for research.

Learning Is a Personal Process

Institutions such as museums are part of the community infrastructure (Falk, 2001a). They are resources to which members of the public have access on a continuous basis and they may choose whether or not to visit and take advantage of the learning experiences that are available. There is a variety of museums and each offers educational opportunities associated with its purpose. The educational opportunities in an aquarium, for example, differ from those in an art gallery or an anthropological museum, although typically, all have displays or exhibitions and aids to their interpretation, such as signage and docents. However, the extent to which visitors choose to engage with, and perhaps learn from, these educational opportunities is their decision. Visitors come with their own agenda, and their level of engagement with exhibits and displays is their choice. Thus, each visitor will construct his or her own pattern of engagement and each will have a uniquely personal learning experience.

A learning experience requires engagement, some mental, physical, or social activity on the part of the learner. Meaning is made from that experience, as Silverman (1995) reminds us, "through a constant process of remembering and connecting" (p. 162). A person's past experiences—be they cognitive, affective, behavioral, social, or cultural—will help to structure the new learning in personal ways. We cannot necessarily *see* that learning has occurred, that new knowledge is gained, a different opinion is held, or there is a disposition to modify behavior, for example. Rather, learning is observable in an individual's actions, that is, what that person does or says. As we shall see, this creates particular challenges for researchers.

Learning Is Contextualized

Although learning is personal, it is not an isolated process. Learning experiences in museums involve people interacting with other people, or with objects and artifacts, in some kind of learning environment, and using socially constructed language. Falk and Dierking (1992, 2000) proposed that three contexts—the personal, social, and physical—interact to produce the nature and outcomes of the visitor's museum experience. We have alluded already to the personal context, which includes the visitor's own background, his or her previous experiences, interests, social skills, and current understandings about the information on display. The social-cultural context refers to other people and the visitor's interaction with them as well as the social and cultural features associated with the artifacts and exhibits. The physical context comprises the physical aspects of the environment of the museum visit, including the architectural features, exhibition layout, the exhibits, their labels, and so on.

This Interactive Experience Model was an important milestone in museum research because it established the need for researchers to take account not only of what happens during a museum visit but also where it happens and with whom. But even more important was the realization that things do not just happen in a context; the context is part of what is happening. An example will clarify. Although not written for this purpose, Rand's (2001) personal account of her rafting trip through the Grand Canyon illustrated how the interaction of all three contexts created her experience. Rand went "along for the ride" but soon found that simply being there did not ensure enjoyment or understanding of the physical environment. Although the Grand Canyon was stunning, Rand found that "until my needs were met, I couldn't learn, I couldn't appreciate, I couldn't turn my attention to higher things: the kind of things a Canyon trip was supposed to be about" (p. 12). Rand's needs were personal, in terms of her comfort, safe access to the scenery, and knowing what to expect, but also social, in that she needed to feel welcomed and included in the excursion group. Once these needs were satisfied, she also needed interpretation of the physical environment in ways that were interesting to her. A museum professional, Rand wrote her account as a parallel to a museum visit, leading her to suggest a Visitor's Bill of Rights, aimed at making visitors feel welcome and providing the context for learning in a range of individually appealing ways.

Feeling welcome is important. Visitors who feel intimidated by the number or intellectual tone of the exhibits, the noise level, an unfriendly physical layout, or apparently aloof attendants will be less motivated to learn than those who feel free to do as they wish. Csikszentmihalyi and Hermanson (1995) discussed motivation in museum settings, pointing out that intrinsically motivated people obtain reward in the pleasure of an activity, "freely expressing themselves by doing what interests them" (p. 68). *Choosing* to visit a museum is a purposeful, intrinsically motivated act—a desire to view a new exhibition, perhaps, or to have an outing with friends and/or family. *Being taken* to a museum provides a different set of circumstances, often accompanied by different kinds of motivation and expectations about the visit. Students on a school excursion will have a different experience than children in a family group, for example. Not only will they have, and will expect to have, different kinds of social interactions, but students often have an imposed structure, such as worksheets or a tour guide.

All of these variables become part of the visit experience—where visitors are, how comfortable they feel, how and with whom they choose to interact, and the nature of their interaction. The personal circumstances of a visit—people's needs, interests, and expectations—all help to contextualize what is learned.

Learning Takes Time

When people see something new that interests them, they try to make sense of it in terms of what they already know or have perceived. Visitors learn during a museum visit when they remember previous information and experiences and connect them with new information and ideas evoked by the exhibits, enabling them to construct new understanding or a different way of thinking or acting. Learning is change and change is not instant. It requires time for reflection—the process that enables us to link new ideas and information with old, to weigh and consider, to deconstruct and reconstruct our mental ideas in order to assimilate and integrate our experiences into new ways of understanding, thinking, and acting. Of course, we have all experienced those sudden flashes of insight depicted by cartoonists as illuminated light bulbs. But these "aha!" experiences simply demonstrate the cumulative nature of learning; the sudden clarity of things falling into place couldn't happen unless those "things" were already part of our mental structures. Current learning can be considered dependent on previous learning or understanding and the basis for building further learning at a later time. Thus, learning is cumulative and iterative; it is an ongoing process rather than a single event.

The cumulative nature of learning means that the significant impact of a museum visit will very often occur not at the museum but sometime later. Visitors show evidence of learning from their museum experience over time, as changes develop in attitudes, behavior, and knowledge. In particular, because learning is contextualized, if the visit is to have any long-term impact, then time is required to allow that learning to find relevance, that is, to be transferred from the context of the museum to other contexts in the visitor's life situations. Falk and Dierking (2000) framed the Interactive Experience Model as the Contextual Model of Learning in a way that emphasized the importance of time as an essential element in properly addressing how learning is shaped by the interacting personal, socialcultural, and physical contexts. An illustration comes from research on visitors' learning from interactive exhibits (Falk, Scott, Dierking, Rennie, & Cohen Jones, 2004), which demonstrated that outcomes in the longer term may be different, and not predictable, from short-term outcomes. Affective and social outcomes were reflected more prominently in the long-term outcomes, suggesting that outcomes "mature" in the context of visitors' subsequent activities. Visit experiences remain as a collection of memories to be awakened by new experiences throughout the visitor's lifetime.

IMPLICATIONS FOR RESEARCH

Learning is a personal process, it is contextualized, and it takes time. These characteristics of learning have long been recognized and their significance has been consistently reinforced in research on learning from museums. However, we believe that they have yet to be consistently well addressed in research, particularly in combination. Although individual research studies may illuminate aspects of one or more of these characteristics, a holistic picture of learning from, and the impact of, museums will only emerge with recognition of all three. We believe good progress is being made toward this end. In 2003, the National Association for Research in Science Teaching endorsed the "Policy Statement of the 'Informal Science Education' Ad Hoc Committee" (Dierking, Falk, Rennie, Anderson, & Ellenbogen, 2003). Although couched in terms of science education, this statement is a much broader document. Similarly, when Rennie, Feher, Dierking, and Falk (2003) expanded the ideas in the policy statement, particularly in terms of directions for further research, their discussion was applicable to learning in any free-choice setting, not only those where there are learning opportunities in science. Here, we wish to complement their effort by exploring the three characteristics of learning and their implications for research design and methods of data collection.

Learning Is a Personal Process

The personal nature of learning focuses attention on the learner as an individual who has a unique visit experience. We noted that learning requires mental, physical, or social engagement by the learner, and that learning outcomes may have any combination of cognitive, affective, behavioral, or social aspects. Further, because we cannot observe directly what learners are thinking, gains in knowledge or understanding, or an attitude change, are observable only in what learners say or do. These features have several implications for research into learning outcomes, which we explore under two headings—the need to "see" the visit and its impact "through the eyes" of the visitor, and the need to consider multiple outcomes.

Seeing Through the Eyes of the Visitor

Visitors must be involved in the research process, not merely observed from a distance, because there is a sizable inferential gap between observing and interpreting what is observed. Seeing through the eyes of the visitor means that, at some stage, data must be collected *from* the visitor and this requires self-report data, or recording what visitors both say and do. Traditionally, interviews and questionnaires have been the main techniques for collecting self-report data and they will continue to be employed, but increasingly, innovative variations that capture more effectively the diversity of visitors' ideas and mental processes are being used. For example, Personal Meaning Mapping (Falk, Moussouri, & Coulson, 1998) is a flexible, interview-based technique that can elicit both cognitive and affective ideas, using visitors' own words. This technique is designed to cope with repeated administrations to demonstrate change over time.

Care must be taken in preparing data-gathering instruments to ensure that their meaning is clear to visitors. Involving visitors in the construction of measures is effective. For example, Johnston (1999) developed two survey instruments designed to measure short- and longer-term impacts of a visit based on written responses of visitors to a postvisit questionnaire. This enabled him to use the words and phrasing chosen by visitors themselves, creating items that were meaningful and relevant to them. Similarly, Rennie and Williams (2002) built their survey about scientific literacy on focus group discussions with visitors. By analyzing how visitors expressed their understandings (and misunderstandings) about science, Rennie and Williams were able to structure their items using terms likely to be familiar to visitors. Both of these studies used extensive field-testing of the instruments to ensure they were unambiguous and "user friendly."

Recording visitors' conversations is a well-used technique for data collection, but for valid interpretation, visitors' actions must be linked to their talk. New technology is helping here. For example, Leinhardt, Knutson, and Crowley (2003) equipped willing visitors with wireless microphones and recorded conversations on minidisks that could be digitally marked later to correspond with visitors' activities noted by an observer. This allowed visitors to conduct their tour uninterrupted, simultaneously providing the researchers with a comprehensive record of conversations and the context of where they happened. New electronic surveillance techniques and "smart" computers, developed for quite different purposes, are creating new possibilities for data-collection techniques.

In order to understand the impact on an individual visitor, researchers must take into account information *about* the visitor, including the purpose for visiting. Roberts (1997) drew on the research of others to begin what she described as "a taxonomy" of purposes and experiences that visitors sought at museums. She included social interaction, reminiscence, fantasies, personal involvement, and restoration (to relax and recharge). Roberts did not discount intellectual curiosity; rather she was demonstrating that "we are seeing only the tip of the iceberg when it comes to understanding what visitors make of museums" (p. 138).

At Smithsonian Institution Museums, Pekarik, Doering, and Karns (1999) accumulated data from nearly 3,000 visitors to explore their views of a satisfying experience. Building from an empirical base of conversations with museumgoers, Pekarik et al. compiled a list of experiences under four headings: objects, cognitive, introspective, and social experience. In studies at nine museums, entrance and exit interviews with representative samples of visitors focused on this list. At the entrance, visitors were asked what experiences they were looking forward to, and at the exit, what experiences they found most satisfying. Pekarik et al. found differences between which kinds of experiences were most satisfying, according to age, sex, and new or repeat visitors. Thus we see how the personal motives of the visitor are an important antecedent to the visit, affecting what happens at the museum.

Capturing Multiple Outcomes

The different motives for a museum visit, and visitors' different experiences, ensure that the outcomes of the visit are likely to be multiple, rather than singular. Even a visit for an ostensibly information-seeking purpose may have affective or sociocultural outcomes, often unintended by the exhibition designer. This means that researchers must be alert for more than cognitive outcomes, and that a wider range of outcome measures must be employed than often has been the case. In the past, much research examined cognitive outcomes, seeking to measure knowledge gained through a visit.

In the context of science centers, and following the ideas of St. John and Perry (1993), Rennie and Williams (2002) have argued that, rather than try to measure bits of scientific knowledge that visitors might learn, researchers should investigate whether their experience has helped visitors to think differently about science. Rennie and Williams developed an instrument that included items about understanding the nature of science and also visitors' confidence and their thinking about the exhibits. They also interviewed visitors to capture in a more personal way their ideas about science.

If researchers are interested in cognitive outcomes from specific exhibits, an important research tool can be a knowledge hierarchy (Perry, 1993), which provides an incremental series of levels of understanding about the concept(s) portrayed in an exhibit. Once developed through "careful examination of the exhibit, discussion with exhibit developers, and in-depth interviews with visitors to the institution" (Perry, 1993, p. 74), a knowledge hierarchy has use in exhibit evaluation as well as visitor research. We believe that, for cognitive outcomes, the potential of the knowledge hierarchy has yet to be fully explored.

Considerable challenges remain in measuring noncognitive outcomes from museum visits. McCrory (2002) listed potential learning outcomes under five domains: cognitive, affective, conative, behavioral, and social. This classification was based on work with teachers whose classes had visited a science center, but the work needs further exploration with other visitors to determine its effectiveness. Another approach was used by Leinhardt et al. (2003), who devised a model of learning as "conversational elaboration," or participation in conversation, and their coding of data across seven exhibitions demonstrated it has explanatory power. Like Leinhardt et al., we also think "an equally important part [of exploring learning] is the development of general conceptual structures that [allow] us to move from meaningful moments of authentic conversation in museums to aggregate quantitative interpretation" (p. 30).

We do not yet fully understand the range of outcomes that indicate the impact of museums. We argue for increasing research interest in outcomes that are salient in an individual's daily life, outcomes that are more likely to be concerned with behavioral and attitudinal changes, such as altered values or opinions. In emphasizing that impacts of the visit are unique and individual, we do not wish to divert attention from research into overall trends. That is important because, ultimately, we seek better understanding of the big picture about how free-choice institutions have an impact on the lives of the public. We are simply arguing that until we have more data about the range of individual outcomes, we are likely to understate the impact of a museum visit.

Learning Is Contextualized

Falk and Dierking (2002) suggested that "Learning begins with the individual. Learning involves others. Learning takes place somewhere" (p. 36). These three statements, simple but not simplistic, capture context—the personal, social, and physical contexts that shape the learning experience. On the one hand, the distinction is artificial because the three contexts overlap and interact. On the other hand, the distinction is helpful because it draws attention to the breadth of contexts, and assists researchers to take a broader view of the visit experience. Twelve factors that influence learning have been teased out as part of the Contextual Model of Learning (see Falk & Dierking, 2000; Falk & Storksdieck, 2005). Together, these factors emphasize the complexity of the research task and, at the same time, they remind us of the broad parameters a full investigation of the visit's impact must embrace.

Much early research on learning from a museum visit focused on aspects of the physical context, such as the location, the exhibits, and the tools and artifacts intended to aid their interpretation. Research that failed to account for visitors as active agents in the visit experience provided only a partial view of outcomes. As the significance of the personal and social contexts of the visit experience gained recognition, research questions began to address this aspect. Visitor research in the 1980s made significant advances in understanding the social behaviors of groups, particularly families, in museums. The research focus on what group members do also includes what they say and how they scaffold each other's learning. Whereas earlier research might have considered only whether docents or interpreters were present or not, research questions now ask about the role they play as they interact with visitors. It is difficult to overestimate the significance of the social context. Even for a solitary visitor, watching and listening to others socially mediates the visit experience.

The implications of these changes in research directions are clear. Understanding the contextual nature of learning requires comprehensive research designs and a range of measurement techniques. Falk and Storksdieck (2005) broke new ground in making the first attempt to account for all 12 factors in the Contextual Model of Learning with visitors to a science center exhibition. Using interviews and observational and behavioral measures before, during, and at the end of a visit, these researchers accounted for 11 factors (the 12th must be measured longitudinally and results are not yet published). Their effort has made a significant contribution to unraveling the contextual complexities of the visit outcomes, and they have demonstrated empirically the multidimensional nature of the visitor's experience.

Research Design

Different kinds of research questions require different kinds of research designs. Nearly half a century ago, Campbell and Fiske (1959) advocated a multitrait, multimethod approach to the validation of research findings; in other words, when several traits are measured by several methods, the relationships among the results will provide both divergent and convergent tests of validity of the outcomes. These researchers wrote in the context of improving psychological measurement, but the foundations were laid for mixed-method research designs (Greene, Caracelli, & Graham, 1989; Johnson & Onwuegbuzie, 2004) and the concepts of methodological and data triangulation (Mathison, 1988) to seek both convergence and divergence in data analysis and research findings. Mixed-method research designs employ a range of methods of enquiry, both quantitative and qualitative, that together produce complementarity and triangulation to better understand the complexity of the research situation. Increasingly, research in the museum environment has employed a range of methods to answer particular research questions (see, for example, Soren, 1995), and we encourage more researchers to accept the challenge of designing their research with this in mind.

The ontological frameworks for the design of the research are also important because they provide the theoretical underpinnings for synthesizing the research findings and their interpretation. Rennie and Stocklmayer (2003) pointed out that museum research was eclectic, garnering theories about learning and research methods from a range of disciplines. In the 1980s, constructivism (the notion that learners construct their own understanding based on prior learning) blossomed in educational research and its role was recognized in the museum context (Hein, 1998, 2005; Roschelle, 1995). It remains fundamental, as our discussion of learning has demonstrated. During the past decade, the Museum Learning Collaborative carried out significant research framed in sociocultural theory (Leinhardt, Crowley, & Knutson, 2002; Leinhardt et al., 2003; Schauble et al., 1997) working particularly with data from visitor conversations, as already noted. Such research highlights the social context of the museum experience, and also recognizes the personal and physical contexts by collecting data not only about their conversation but also about how participants are simultaneously dealing with the learning environment. Leinhardt et al. (2003) described how their data were being used to develop general conceptual structures and move the theoretical agenda forward. Perhaps now the "exporting" of theories from museum research to research in education and psychology has begun to address the "trade imbalance" referred to by Paris and Ash (2000), who drew attention to the importing of educational and psychological theories into museum research.

Measurement Matters

Dealing adequately with context in data collection requires multiple measures, as demonstrated in the Falk and Storksdieck (2005) study. Data must be collected not only about the visitors themselves but also about where the visitors are, what they do, and with whom they interact. A variety of data-collection techniques will be required to describe fully the visit experience, and the research design must ensure complementarity and triangulation of data for trustworthy interpretation. Descriptive data collected at the time of the visit can include field observations as well as various electronic means of recording visitors' conversations and activity. Other data from visitors can be collected by the usual means, such as interviews or pen-and-paper surveys (see Hein, 1998, for an overview). Interestingly, the opinions of floor staff are rarely sought, yet they are "professional people watchers." We have found them to be very insightful about visitors' needs and actions (Johnston & Rennie, 1995; Rennie & Johnston, 1997).

An important consideration in research with people is the potential for reactivity in the process of measurement. How does one collect data in ways that do not change the behavior of the visitor? This problem has long been recognized. In 1966, for example, Webb, Campbell, Schwartz, and

Sechrest wrote about unobtrusive measures or ways to collect data without alerting those about whom the data are collected. In the museum context, they gave examples of "physical traces," such as floor wear and the number and height of nose prints on the glass of exhibit cases, as unobtrusive ways to determine how much attention exhibits were attracting. Although these measures tell which exhibits are popular and which are not, they have little to say about the impact on the visitor. Lucas, McManus, and Thomas (1986) emphasized the importance of preserving the "informal learning" context for valid results and discussed some of the difficulties arising from using measures such as observation and recording. Barriault (1999), Borun, Chambers, and Cleghorn (1996), and Griffin (1999) have provided lists or frameworks of observational ways to identify whether learning is occurring. We need more research using these noninterventionist methods to determine how valid they can be and under what conditions they are most effective.

Attention must also be given to the ethical issues relating to research. There is a delicate balance between observing or recording visitors' behavior and infringing upon their personal rights, especially as we move to newer, less intrusive, technologies for data collection. Researchers cannot ignore the possible negative implications and must ensure that visitors are informed about the research process. We know of more than one researcher whose observational activities have led concerned onlookers to alert security personnel!

Learning Takes Time

Allowing time for learning means that the investigation of impact must be ongoing, not a once-only incursion during the visitor's museum experience. Research designs must include the opportunity for collecting data in a longitudinal way, and longitudinal studies require measurement over time, ideally before a visit as well as during and after. This was realized implicitly in 1929, when a series of studies began on the educational impact of the museum visit on children (Melton, Feldman, & Mason, 1936). Melton et al. pointed out that major factors needing investigation were the methods used to prepare children for the visit, the nature of instruction during the visit, and the educational accomplishments afterwards. Melton et al. used tightly controlled approaches to data collection (as well as methods of instruction), as befitted the research thinking of the period. Now we recognize that measures must be flexible and wide ranging, to capture both expected and unexpected outcomes for the visitor. Because visitors are all different, the outcomes of a visit and subsequent impacts on visitors' lives can be both diverse and unpredictable. Interviews or written means of data collection that are very structured risk missing important variations in how visitors think, feel, and act as a result of their visit and how that visit has an impact on their lives.

It is evident that longitudinal research takes time, but it is the time it takes that is the greatest barrier to carrying it out. In practical terms, funds for research are usually limited and results are needed sooner not later. In research terms, the longer the study, the greater the difficulty in retaining the sample; further, the potential for reactivity is a continuing problem for longitudinal researchers. The act of collecting data at one time may cue the visitor for data collection at a later time, as well as influence their subsequent behavior. Postvisit interviews, particularly with stimulated recall through photographs or video recordings, have the potential to change the impact of the visitors' experiences by requesting visitors to think and talk about the experience in ways they might not otherwise do. Although the actual visit experience might be better understood, the researcher's intrusion may affect the longer-term impact, and it becomes very difficult to disentangle possible researcher effects.

We see that time is a particularly tricky variable in research on learning from the museum visit. Realistically, we cannot shadow visitors for life, so an integrated series of "snapshots" based on a range of data-gathering techniques is our best chance of coping with time. Ellenbogen's (2003) research, following families before, during, and after their museum visits, made an important contribution here, because she was able to study a sequence in these families' lives, revealing the interconnectedness between the visits and other family activities.

CONCLUDING COMMENTS

It is important to remember that, in terms of the visitor's lifetime, the museum visit itself is a fleeting event. Visits are measured in minutes or hours, not days. A visit, or even repeated visits to a museum or museums, is just a tiny part of an individual's total learning experience. We can illustrate this using an age-old metaphor: even the longest museum visit is like a tiny thread woven into the tapestry of the visitor's life experiences,[1] linked directly or indirectly to the other threads. Any learning that occurred during the museum visit becomes part of the weft and warp of the visitor's total lifetime experiences. How does a researcher identify that tiny thread and map accurately its impact on the total tapestry? To return to our example of the fictitious aquarium visitors, the cuttlefish watcher's "tapestry thread" plays a minor role; removing it from the tapestry may have little effect. In contrast, pulling the "thread" of the sea grass exhibit viewer may begin to unravel the whole tapestry, because that tiny thread is the basis of a lifetime commitment to conservation, reflected in how that person's life is lived.

As researchers, we want to ask questions like: How much and what type of learning results from a museum visit? How much measurable impact is there likely to be later in a visitor's life? We must understand that the answers are not simple. They depend on the visitor, on the context of the visit, and what has happened or will happen over time, that is, before and after the museum visit. The most satisfactory answers will come from research questions that are realistic in terms of coping with the variability of outcomes, from approaches to data collection that account for those multiple outcomes by using measures that "see through the eyes of the visitor," and from research designs that are sufficiently broad based to capture a range of impacts, some trivial, some amazingly complex, and some that become evident well after the visit. We remain optimistic. Research into learning in museums has always been difficult but progress is being made. Evolving conceptual models of visitors' experiences are developing appropriate, alternative methods of research. This array of methods is building our understanding of free-choice learning by triangulating with, and adding to, an accumulating database. We believe that further consideration of the personal nature of learning, the context of that learning, and the time taken to process and assimilate new information, provides the overarching principles for continuing research into learning in museums and investigating the impacts they have on people's lives.

ENDNOTES

An earlier version of this chapter was published in *Science Education*, 88 (suppl.1): S34–S47, 2004, © Wiley Periodicals, Inc. Used with permission of John Wiley & Sons, Inc.

1. Tapestry as a metaphor can be traced back at least to the Greek Theocritus, born in the third century BC, who composed pastoral poems. Its consistent use is not surprising, as the main purpose of tapestries was to tell stories about people's activities and lives.

II

ENGAGING AUDIENCES IN MEANINGFUL LEARNING

6

Envisioning the Customized Museum: An Agenda to Guide Reflective Practice and Research

Mary Ellen Munley, Randy C. Roberts, Barbara Soren, and Jeff Hayward

Customization is a ubiquitous feature of contemporary life. One person's double-shot latte with soy milk, no sugar, and a dash of caramel flavoring may not be another's "cup of tea"—but not to worry, hand-picked black tea infused with maple syrup (grade 2 dark) is coming right up. The boomer generation, true to its values of self-reliance, entitlement, free choice, and nonconformity (Gillon, 2004), ushered in the era of individual choice. This deep-seated belief in choice and an equally strong disdain when faced with a situation that does not recognize individual preferences is at the heart of the classic diner scene in the 1970 film *Five Easy Pieces*.

Robert Dupea, played by Jack Nicholson, attempts to order a plain omelet, no potatoes on the plate, a cup of coffee, and a side order of wheat toast. The waitress informs him that there are no substitutes to the items on the menu, and on the menu, plain omelets come with cottage fries and rolls—no toast. She explains politely, "I'm sorry, we don't have any side orders of toast . . . an English muffin or a coffee roll." Responding to the inherent ridiculousness of that statement, Dupea goes to work . . .

Dupea: What do you mean you don't make side orders of toast? You make sandwiches, don't you?
Waitress: Would you like to talk to the manager?
Dupea: You've got bread and a toaster of some kind?
Waitress: I don't make the rules.
Dupea: OK, I'll make it as easy for you as I can. I'd like an omelet,

plain, and a chicken salad sandwich on wheat toast, no mayonnaise, no butter, no lettuce. And a cup of coffee.

Waitress: A number two, chicken sal san, hold the butter, the lettuce and the mayonnaise. And a cup of coffee. Anything else?

Dupea: Yeah. Now all you have to do is hold the chicken, bring me the toast, give me a check for the chicken salad sandwich, and you haven't broken any rules.

Today, some 40 years after that memorable plea for customization captured the zeitgeist of a generation, it's no longer necessary to muster quite as much imagination or antiestablishment bravado to get your way. Customization has swept the nation. Generations X and Y are also on the bandwagon bringing TiVo, MP3 players, and customized web searches to the contemporary lifestyle of getting just what is wanted precisely when it is wanted. Customization is a significant cultural development. Rayport's (2005) description of it and his call for demand-side innovation was a *Harvard Business Review* breakthrough idea for 2005, and international trend watchers like Gerald Celente's Trends Research Institute identify customization as one of the most established and important-to-watch trends of our time.

Customization has exploded in retail and service industries—a perfect match to businesses where "give the customer what he wants" is a sacred mantra. Does the undeniable success of customization in business and its pervasiveness in our society suggest that it is an equally good fit for the world of free-choice learning? Will visitors and other museum patrons expect more individualized attention and services? What happens when the items to be customized are not products and services like jeans, ice cream, and maps to use on a vacation trip? What if the aim is to customize experiences with art, history, or science? What about public education and cultural heritage? Should they be customized? And if these experiences were given over to customization, what happens to treasured ideas like intended messages, shared experiences, belongingness, and collective cultural identity?

Customization in the world of cultural institutions and public education gets complicated. This chapter focuses on how the idea of customization can be of use in one type of informal, public education environment—the museum. It includes ideas, definitions, and questions to guide a thoughtful exploration of the powerful and ubiquitous idea of customization and suggests avenues for research to guide investigating what about the idea of customization, as introduced in business practices, is a good fit for museums—and what is not.

WHAT IS CUSTOMIZATION?

There is little doubt that customization is a business and societal trend that is—and should be—migrating into the museum world. Nonsequential and multitrack audio guides, "Ask Me" floor staff, a selection of family or adult programs, websites with access to images and information not on public display, and content provided for download on visitors' personal MP3 players are all evidence of museums responding—whether consciously or not—to the customization trend.

Customization, connected first to consumer goods and more recently to services and experiences, has been a subject of increased study among business and management scholars. Mass customization (Pine, 1993; Pine, Peppers, & Rogers, 1995), the concept of providing personalized products at affordable prices, has been described by Pine (1993) as a paradigm shift requiring businesses to progress from a structure and philosophy that supports mass production to one that supports individualized service. Pine's approach is based on the underlying belief that "customers do not want choice; they want exactly what they want" (Pine, 1998).

The need for this rethinking of core approaches in business arises from an increasing fragmentation of the mass market. Not only is there more diversity within the market, but the individual needs and desires of each customer are more prone to shifts and changes. Goldsmith and Freiden (2004) identify two management trends behind the growth in what they call "one-to-one" marketing: (1) more attention to customer value and satisfaction, and (2) new technology enabling greater knowledge about consumer behavior.

In business, advances in information technology and manufacturing processes are at the heart of facilitating flexible and cost-effective ways to meet customer needs in a more individualized way (Gilmore & Pine, 1997; Mok, Stutts, & Wong, 2000). In addition, the increasing ubiquity of technology in the lives of consumers is a major force in shifting customer expectations in the marketplace (Hart, 1995).

Steven Weil (2002) observed a similar paradigm shift and set of trends in museum practice. A set of factors similar to those found in business—a desire to serve a more diverse population, recognition of the different backgrounds and expectations found among audiences, increased attention to customer service and visitor-centeredness, and a realization of the ways technology has affected people's expectations about access to information and services—is alive in the museum world.

So if customization is a good response to these conditions for business, it seems logical to hypothesize its value for museums. Some caution, however, is in order. Determining what museums can learn from customization practices in business needs to be grounded in an analysis of the idea of customization that includes an understanding of the similarities and differences between the domains of business and public service. Just as with the adaptation of any business model to the social sector, it is essential to borrow only those aspects of the model that are a good fit. Moore (1995) and Collins (2005) caution that while business and public service sectors share a dedication to discipline, they are markedly different in terms of motivation (mission) and measures of performance. As Collins (2005, p. 1) puts it, "we must reject the idea—well-intentioned, but dead wrong—that the primary path to greatness in the social sectors is to become 'more like a business.'" The investigation of the best fit of the idea of customization to museums will benefit from this wise counsel to borrow only what fits and make alterations as needed. The challenge for museums, as Falk and Shepherd (2006) articulate, is to craft a definition of customization in the public service arena of museums that draws on features of the business-world definition but also takes into account the distinctive features of museums and their role in communities and the larger society.

Hart (1995) offers two distinct definitions of mass customization for business: the visionary definition and the practical definition. The ultimate (and, he asserts, essentially unachievable) goal—the visionary definition—is "the ability to provide your customers with anything they want profitably, any time they want it, anywhere they want it, any way they want it." Practically, Hart defines customization as "the use of flexible processes and organizational structures to produce varied and often individually customized products and services at the low cost of a standardized, mass-production system." Thus, for business, customization helps improve the bottom line. The *motivation* for adopting a customization paradigm is to increase market share and sales, and generate more revenue *by* giving customers exactly what they want *without* incurring additional costs for production.

The motivations and end results of customization are different for businesses and cultural institutions. Businesses want many satisfied customers in order to achieve a better profit margin, while cultural institutions—like museums—want many satisfied visitors and patrons in order to realize their missions and provide better public service. "The customer is always right" and "give them what they want" are the most essential guiding principles for a business working to achieve its goals through customization. Those principles are not the main forces that guide cultural institutions. Museums, for instance, are guided by the collections, content, messages, and perspectives they have to share—even when those messages may make people uncomfortable by examining prejudices or shortcomings or by having them explore unknown territory. For museums, the value of customization is that it can increase their public value. The *motivation* for adopting a customization paradigm in a museum is to increase the number and kinds of people who benefit from a museum's offerings and to deepen their engagement with art, science, nature, and history *by* providing them with a variety of ways to engage with content and ideas *without* violating the values and ethical standards that guide the institution.

WHAT CUSTOMIZATION LOOKS LIKE

It's always wise to begin a new investigation by questioning assumptions and asking very basic questions. And so research about the efficacy of customization in museums might start with these fundamental questions: Are museums already perceived as customized, free-choice environments by their audiences? What are the distinguishing features of museums that are perceived as highly customized and what are those that are not? Do visitors want to customize their museum experiences? If so, what do they wish they could have or do that is not available to them now? Do museum staff, administrators, and trustees understand and believe in the idea of customization? What are examples of customization that are already in practice? And what effects are they having on visitors' experiences and the community's perception of the value of the museum? Research designed to investigate these fundamental questions would certainly advance our understanding of customization and its defining features.

In addition to examining fundamental questions, another approach to advancing knowledge is to encourage museums to experiment with customization and to conduct evaluation and audience research studies. Gilmore and Pine (1997) describe a flexible set of approaches to customization that could be applied either individually or in concert with each other, depending on the situation. The four approaches they describe for business could easily provide a framework for experimentation with customization in a museum:

- *Collaborative customization* involves dialogue with customers to help identify their needs, identify what you can offer that fulfills those needs, and create a product to fill customer needs and desires.
- *Adaptive customization* offers a standard product that is designed so users can customize it themselves.
- *Cosmetic customization* offers a standard product that is packaged specially for the individual.
- *Transparent customization* provides individual customers with unique goods or services without letting them know explicitly that those products and services have been customized for them. This is accomplished by observing customer behavior without direct interaction and then customizing offerings within a standard package.

Examples of each of these approaches can be identified in the marketplace today. Standard desk chairs that can be adjusted by customers to fit their specific body shape are an example of adaptive customization. Dell Computer with its build-to-order model uses the collaborative approach by assembling products to meet each customer's specifications. The Planters Company recently retooled its plant in Suffolk, Virginia, to enable cosmetic customization by creating more variety in the size and design of packaging to better meet the individual needs of their retailers (Mok et al., 2000). The Ritz Carlton and many other hotels observe and record preferences of their customers and presciently fulfill these preferences on return visits (Pine, 2002). This is an example of transparent customization, a strategy that abounds in e-commerce and other Internet applications.

Gilmore and Pine's four types of customization provide a robust framework for crafting customized products and experiences in museums. Museums engaged in front-end evaluation as they prepare new exhibitions are leading the way with collaborative customization. What other aspects of a museum's offerings lend themselves to collaborative customization? Audio tours that allow the user to select the sequence of stops is a good example of adaptive customization—the total set of recordings is standard, but the way they are accessed, and thus the experience, is varied for each visitor. Cosmetic customization may seem trivial, but it may in fact make a difference in who is attracted to an offering. Thinking in terms of cosmetic customization, an adult lecture series could be promoted to an audience of teachers simply by marketing to them and providing professional development credits. Examples of transparent customization are rare in museums, but they do exist. Education departments that keep records of program attendance and market a summer camp program to families who enrolled in family and children programs in the last two years are exhibiting intuitive knowledge of the promise of transparent customization. Development departments are expert transparent customizers. Matching donor interests with museum activities is second nature to them. What might be learned from the special attention museums give their most prized financial supporters that would improve their capacity to meet their educational mission for broader audiences?

CUSTOMIZATION AND LEARNING

The idea of customization is a superb fit with best practices in museum education. The foundations of today's experiential and inquiry-based approaches to program and exhibition design can be traced to the progressive education philosophy of Dewey (1938) and the extraordinary work of matching a museum to the needs of its community championed by John Cotton Dana (1917). People are active, not passive, learners, they argued, and Dewey and Dana created learning environments at the Barnes Collection and Newark Museum, respectively, that remain to this day as singularly serious and successful examples of museums responding to the desires and needs of individual learners and communities while staying true to their values and missions and the integrity of a thoughtful search for knowledge and understanding.

Dewey, Dana, and today's constructivists (e.g., Hein, 1998; Jeffery-Clay, 1998) describe museums as ideal learning environments because they allow visitors to explore freely, move at their own pace, interact and share experiences with groups, and examine and expand their own understanding. Sounds a lot like customization!

Yet, when examined closely, how many museums are taking full advantage of their capacity to facilitate meaningful interactions between their resources, the people who view them, and the communities that house them? The opportunity to move freely from one gallery to another or to attend to one section of a display and not another does provide a visitor with choices, but it is not customization.

It is quite possible that a customization framework—when more fully explored—could release museums from their relatively simplified notions of customization and, in fact, result in ways to work much harder and smarter to support what Hooper-Greenhill (2000) describes as the process by which visitors "search for meaning, look for patterns, try to invest their experiences with significance." Carr (1999) believes that an important outcome of this more customized approach to learning is that museum visitors ask good questions. "It seems to me," Carr postulates, "that our most important task as educators is to assist human beings to craft their cognitive lives through the expression and pursuit of the unknown. That is: our task is to help people to ask good questions" (p. 3).

Indeed, customization is driven by self-knowledge and the capacity—and permission—to ask questions. This suggests that practice, research, and evaluation drawn from the perspective of customization might focus on questions, rather than answers. What inspires people to reflect on what they know? When asked, what questions do people have about a museum? An exhibition? An artifact? Is there a role for the museum in assisting people in crafting really good questions? Do some questions inspire more exploration then others?

Letting questions, not answers, take the position of prominence is a potentially very powerful result of adopting a customization framework, and it is a comfortable fit with efforts being made in scores of museums to shift to inquiry-based, visitor-centered experiences where the role of the institution changes (see table 6.1).

Equally important, customization changes the role of the visitor as well (see table 6.2).

Table 6.1. Customization Requires Changes in the Role of the Museum

From	*To*
Authority figure	Partner
Delivering knowledge	Facilitating learning
Designing static exhibits	Crafting flexible learning environments
Providing information	Posing questions
Presenting programs	Creating learning tools and providing forums
Finite set of messages and meanings	Multiple messages and meanings

These fundamental changes for museums and visitors result in what Vergeront (2004) calls designing for the visitors, not designing by the museum. Although the distinction may seem insignificant or perhaps even impossible to accomplish, it does signal an important change in guiding principle. It is the difference, for instance, between organizing learning materials for schools around the collections and galleries of the museum's organization, rather than according to the organization of the school's grade levels and curriculum. It is the difference between the museum writing a history trail guide for visitors to follow and a publication like "Walk a Mile in Her Shoes: A Field Guide for Creating Women's History Trails" (2001), a program guide by the Women of the West Museum encouraging people to create their own trails.

At the bolder end of the customization continuum, there is a great deal to explore about how people learn and how to design environments that accommodate different needs. The cutting edge is already being explored. The Learning Shoe, for instance, made its debut in a museum—the San Francisco Museum of Modern Art. Chips in the sole of the shoe conceived by the innovative designer Yves Bhar gather data on the shape of a person's foot and the way he walks. The person circles the block a few times in the yellow trainer shoes and returns to the shoe store where the chips are removed and used to design a new pair of ergonomically customized shoes.

The Learning Shoe is still only a concept product, but research about how the brain processes information and the variety of ways people re-

Table 6.2. Customization Results in Changes in the Behavior of the Visitor

From	*To*
Digesting information	Participating in a "conversation"
Getting answers	Asking questions
No or little control of learning	More control of learning

ceive and process experiences, most frequently understood through application of Gardner's (1993) work on multiple intelligences, provides ample data for experimenting with the creation of flexible learning environments in museums that account for advancing knowledge about how humans learn. The architect Eva Maddox provided a glimpse of a new direction for education environments in her installation as part of *Chicago Architecture Ten Visions* on display at the Art Institute of Chicago in 2004. Working from an understanding of multiple intelligences and advances in access to information, Maddox asserted in her artist's statement that "This new environment reflects the shift from the historical one-way delivery of information to a fully immersive, interactive environment." Museums experimenting with more interactive exhibitions are struggling with similar challenges, but the experiments may not be bold enough once one more fully understands the nature of individual learning and the potential of technological advances.

CUSTOMIZATION AND TECHNOLOGY

Technology is clearly a driving force in the cultural shift toward customization, which makes sense because technologies often ease the way toward personalizing access to information and services. In business, customization does take on a technology-heavy aura. And while technology is also an essential tool for the modern museum, as the previous sections of this chapter argue, it would be a mistake to conflate customization with technology. It would be equally faulty to ignore its usefulness.

In spite of their relative slowness in recognizing the importance of applying technology to individualize customer experiences (Bowen & Filippini-Fantoni, 2004), cultural institutions now appreciate the wealth of potential applications. Across the globe, museums are beginning to actively explore ways of using technology to customize museum experiences, both in physical and virtual spaces. Among the most groundbreaking of these efforts are projects that explore integration of physical space and electronic information systems. Since the early 1990s, ongoing research has examined techniques for merging the physical with the virtual world. A new generation of technologies is emerging that makes widely available the possibility of integrating mobile computing applications into the user's environment (Barton &

Kindberg, 2001). These new technologies and approaches offer a unique opportunity to combine the authenticity of learning from objects with the rich accessibility of e-information.

The electronic guidebook is a physical/virtual approach based on the traditional visitor guide used to gather information in advance of visits and augment information on site (Hsi, 2002; Spasojevic & Kindberg, 2001). The Electronic Guidebook Research Project was initiated in 1998 by the Exploratorium in partnership with Hewlett-Packard Labs and the Concord Consortium. For the project, the museum was set up with technologies created by the CoolTown research program at HP Labs. A variety of sensing devices, like infrared receivers and barcode readers, were integrated with wirelessly networked handheld units. Nomadic users could access electronic services by using their sensors to pick up signals from the physical locations of interest to them (Spasojevic & Kindberg, 2001). In addition to providing more information and wider context, the electronic guidebook technology also was intended to create a more individualized experience.

Galani and Chalmers (2002) have prototyped a system infrastructure that bridges digital and physical visitation to an exhibition and supports interaction between physical and virtual visitors. In their study at The Lighthouse in Glasgow, U.K., Galani and Chalmers used a CyberJacket created by Bristol University, fitted with a device to register the wearer's position within the room, an electronic compass to show her orientation, and cables connecting the devices to a handheld computer that she carried. As the visitor moved around the space, relevant information about the objects and displays with which she was engaged was transmitted to the handheld computer. Other virtual visitors visited the same space on remote laptop computers. The virtual and physical visitors were equipped with audio devices and could discuss their experiences with each other (Galani & Chalmers, 2002).

The Experience Music Project (EMP) in Seattle is also using mobile computing devices called MEGs (Music Experience Guides) to personalize and augment their visitor experience. In addition to serving as an audio guide, the MEG enables visitors to bookmark the objects and exhibits of interest to them. At the end of the visit the bookmarks are uploaded to a main server. Visitors can access EMP's site using the number on their admission ticket and locate a personal web page with information and links based on their bookmarks (Gilmore & Pine, 2002).

Using technology similar to the Experience Music Project, the Visite Plus program at the Cite des Sciences et de l'Industrie in Paris allows visitors to configure a personal profile on a kiosk in the museum. The information is transmitted to a PDA or bar-coded ticket, which is carried through the exhibitions by the visitor, and programmed to activate information and activities based on the user's personal profile. The visitor's path through the exhibitions and a record of his or her activity is automatically saved on a personal web page that is accessible on the museum's website through the number of the bar-coded ticket or PDA. The website includes information and activities about the objects and exhibits in which the visitor indicated interest (Bowen & Filippini-Fantoni, 2004).

CUSTOMIZATION AS PARADIGM SHIFT

Watzlawick (1974) in his classic treatise on change makes a useful observation: "The French proverb according to which the more something changes the more it remains the same," he writes, "is more than a witticism. It is a wonderfully concise expression of the puzzling and paradoxical relationship between persistence and change" (Watzlawick, Weakland, & Fisch, 1974). What is the nature of the change that needs to be happening in museums?

Watzlawick and his colleagues (1974) differentiate two kinds of change. First-order change occurs within a given system that itself remains unchanged. Changing the interpretation and design of exhibitions, improving the readability of labels, and improving standards for collection care are examples of first-order change. Second-order change, on the other hand, changes the system itself. Second-order change questions the premise that things *should be* a certain way—and it is the basic premise, not the way the premise is actualized, that is viewed as the problem in need of change. Customization, with its premise that a museum share control with visitors and patrons—is an example of a change in premise—a second-order change for museums.

Hart (1995) offers a framework for developing a strategy for change that is based on four pillars that need to be examined when considering a shift to mass customization. He describes these four pillars as: customer customization sensitivity, process amenability, competitive environment, and organizational readiness. These "pillars" or decision factors call on those considering adopting customization strategy to ask themselves questions like: Do visitors and patrons care whether they're offered more customization? What resources will it take to convert your operations to accommodate this new approach? Do you know enough about your visitors' needs to craft customized products and services? Do you have the capacity to gather and analyze information about visitors? How would the change position you with respect to your community? And, is your museum ready to take on the organizational changes required to successfully complete this shift in approach? Hart asserts that his "four-pillar" framework can be used to examine an organization's readiness to adopt a mass customization strategy, but the development and implementation of that strategy is unique and cannot be taken from one place and applied to another. "Mass customization," he writes, "is not a one-size-fits-all business concept" (Hart, 1995). Museum self-studies and other research on customization readiness are ripe for investigation and could provide museums with a vehicle for exploring customization, understanding its implications, and determining how it might fit.

Pine (1993) suggests a specific progression for moving from mass production to mass customization: (1) customize services around standard products; (2) create customizable products; (3) provide point of delivery customization; (4) provide quick response; and (5) modularize components. Each of these steps moves away from a mass-production approach and toward a mass-customization approach. Pine's progression suggests an action plan for a museum: What would it look like to customize a permanent exhibition? What would a customized membership program involve? How might a museum assist a visitor arriving at its entrance in outlining a customized experience? The possibilities reveal new horizons for museums, but customization is not a panacea. Issues of feasibility, skill sets, and priority are bound to emerge and will need attention. In addition, there are some noodlesome questions that emerge when discussions of customization move beyond superficial levels.

THE PITFALLS OF CUSTOMIZATION

Adoption of a customization framework needs to be mindful of the fallacy of the superiority of individualization. The same-thing-for-everyone approach is not inherently a bad idea, nor is it to be avoided in all situations. Near-universal adoption of the same computer operating system ensures that a document sent to another computer will open easily and will have the same formatting that it was created with. For many, part of the enjoyment of riding cable cars in San Francisco is knowing that a parent or dear friend had the same experience. There is no disappointment or public outcry because there is only one *Gone with the Wind* or *Guess Who's Coming to Dinner*—or Michelangelo's *David* or van Gogh's *Starry Night.* There is something special—and important to feelings of belongingness and cultural heritage—when people share the experience of the same work of art, cable car, movie, or television program.

The spirit of self-direction inherent in customization can result in more meaningful experiences, but self-direction can also limit experience. It is quite possible that putting decisions in the hands of visitors could limit experimentation, reduce the willingness to wander, and ultimately stifle learning and personal growth. If visitors seek—and get—that which is within their comfort zone (one of the purposes of customization in the business world), what motivates them to leave that zone, to step into unknown and possibly uncomfortable territory (a role most educational institutions embrace)? As museums approach customization, they will need to explore what inspires people to reach beyond what they are already interested in and know.

Increased customization has the potential to decrease social interaction and lead to increased isolation. This is especially the case when customization is realized through the use of technologies that bring all needed information directly to the user and allow little or no interface with real people, real objects, or real issues.

At its essence, museum customization is about sharing control of the dissemination and interpretation of artifacts and information with visitors and other museum patrons. It is not simply about museums' and visitors' choices—it is about museums' and visitors' wishes. Any museum's wishing to expand the audience of people who find genuine value in interacting with its resources would be wise to investigate and experiment with customization. And any person or community wishing to find personal meaning and public value from its museums would be wise to request more customization. If they do, the next few decades look promising indeed.

7

Museums and Cultural Understanding

Gretchen M. Jennings

NEW WINE IN OLD SKINS?

This chapter arises from a dialogue begun at the *In Principle, In Practice* conference held in Annapolis in November 2004. At a general session aimed at identifying important issues in the museum field, several participants spoke of the challenge that creationism, especially in the form of intelligent design, poses to natural history and science museums. Can these museums remain true to their scientific mission and at the same time serve large segments of their audiences that question one of the foundational theories of modern science? A second perspective in this discussion, also related to museums and their missions, concerned the emergence of museums that are questioning long-held practices of collecting, including the care, display, and scholarship of their collections. These museums are reclaiming stories and objects that were once in natural history, art, or anthropology museums and are placing them front and center in institutions representing their own cultural perspective. This chapter will attempt to address these different but related challenges to the museum world of the 21st century:

- In the face of opposition from creationist visitors, how can science and natural history museums remain true to their institutional missions? Should they try to serve all of their constituents, including those who reject important elements of modern science?

- Why is it that, increasingly, the creation narratives of non-Western communities are given respectful display in major museums while the museum field is generally resistant to the requests for "equal time" by proponents of creationism or intelligent design? Is this a principled stance or a hypocritical position?
- What do these two questions say about the nature of the museum itself? To what extent can and should museums remain a kind of embodiment of the Western scientific tradition? And in what cases should these old skins be reshaped to contain new wine?

These are not easy questions, and they demonstrate the issues museums face as they strive to accomplish their missions while also accommodating the points of view of an increasingly diverse set of publics. This chapter proposes to reflect on these questions through the exploration of a number of ideas: the concept of cultural perspective or point of view, the ways in which museums have evolved within the Western perspective, their close alliance with the development of modern science and social science, their connection to Western imperialism and racism, and their emergence in the late 20th and early 21st century as part of a global "cultural flow."

MUSEUMS AND CULTURAL PERSPECTIVE

What Is a Cultural Perspective?

Anthropologist Dorothy Lee once defined culture as a "symbolic system," that is, a system of learned meanings "that transforms the physical reality, *what is there*, into experienced reality." Looking out her window, Lee asks if the trees blocking her view are something she owns and could cut down if she wishes, or are they, as the Dakota people believe, standing peoples with rights to the land? Lee contends that a study of culture allows us to see the variety of viewpoints from which the world can be viewed (Lee, 1959).

I am defining culture here as a self-consistent set of assumptions about how the world works—assumptions about causality, about social relationships, about political and economic structures, about what is living and what is not living, and about how time is measured and experienced. For most of us, culture is like a pair of lenses that are invisible, both to us and to those who share our worldview. We assume that the way we view the world is the way it is, and that everyone views it the same way. Anthropologists call this "ethnocentrism," a perception centered on one's own culture. Awareness that we all wear these cultural lenses, and that it takes intellectual work and conscious effort to remove them or see through them, took some time for humankind to understand.

The Development of "Perspectivity" in the Modern World

Changes in the sciences, the social sciences, and other disciplines have contributed to increased awareness of the phenomenon of culture. And with this awareness has come an understanding of the idea of perspective, or "point of view." For example, both anthropology in the later 19th century and psychology in the 20th century began to explore the existence of radically differing points of view in the human subjects of their respective disciplines. The idea that cultures should be studied on their own terms rather than evaluated against the standards of Western culture began to emerge as a basic tenet in the field of anthropology. Studies in psychology have shown that different people viewing the same stimulus will understand and react to it in different ways (Kuhn, 1970). Developmental psychologists such as Jean Piaget have studied the ways in which our perception and understanding of the world change radically as we develop across the life span. In the field of history the idea that human cultures, or at least Western culture, march inexorably in a single, "progressive" direction has been rejected, and historians are encouraged to study each historical period within its own context.

In the 19th century, biblical studies began to be informed by the fields of history, linguistics, anthropology, and archeology. These studies have led, particularly in the 20th century, to an understanding of Scripture as literature shaped by the customs, language, and values of the period in which it was written. This, in turn, has led to an interpretation of the Bible as a source of truth about God's relationship to human beings and to creation—but not necessarily a factual account of natural or human history. At the same time, there have always been "literalists" who believed "received doctrine" was factually accurate and that interpretation was close to blasphemy. The influential writings of scientist and historian of science Thomas Kuhn proposed that science itself develops within paradigms or systems of meaning that eventually prove unable to explain new findings, and thus are redefined and cast off as new paradigms develop (Kuhn, 1970).

"Western Culture"—A Contested Yet Useful Term

In the late 20th century, anthropologists began to talk not only of cultural perspective but of "cultural flow." As access to news, music, film, art, and other kinds of information becomes global, it is increasingly difficult to talk about cultures as distinct and separate entities. The use of English and the spread of "Western" ideas and values have certainly had an impact on the entire world, but the influence is not all unidirectional. The growing importance of China, India, and the Arab world in the global economic and political scene is already contributing to an ebb and flow of cultural influence. The traditional "Western Civ" course at colleges and universities has come under fire because it places a single culture front and center in the development of world history, and because many scholars question whether one can trace a direct line from Mesopotamia through Greece, Rome, and Western Europe in the development of what has been called Western culture (Birken, 1992). The role of Arab and Muslim cultures in shaping and preserving many aspects of what we call Western traditions today (particularly in the areas of science and mathematics) is being given increasing and long overdue recognition. And scholars argue whether the spread of Western creations like McDonald's or MTV results in the Westernization and homogenization of other cultures or whether world cultures are absorbing and reshaping fast food and rock music to their own models (Breidenbach & Zukrigl, 1999). Although I am fully aware of these issues and at many levels agree, I will continue to use the term "Western" in this chapter because it remains a useful, if imprecise, catch-all for the varied origins of museums today.

MUSEUMS AS EXPRESSIONS OF WESTERN CULTURE

Early Development of Museums

A number of authors have written about the museum as a quintessential Western institution. Museums first developed in Europe as "cabinets of curiosity," displays of wealth and acquisition by aristocrats, explorers, and merchants during the Renaissance and the centuries of European exploration. Their owners placed objets d'art, spoils of war and conquest, and exotic materials collected from all over the world on view in their private homes and palaces for admiration and study by their friends. The institution of the museum has been described as imposing a Western "way of seeing" that transforms all things, whether a giant crab caught during an exploratory expedition or a decorative door from a Nigerian palace, into objects of visual interest, into a Western definition of art (Alpers, 1991). In the late 19th and early 20th centuries, when the modern museum truly developed, it was part of a conscious effort by governments and leaders of society to provide a civilizing influence on the wider society—in other words, to teach the fundamentals of Western culture to the masses. In categorizing and organizing displays of different "exotic" cultures, and the flora and fauna of both home and foreign lands, natural history museums and world expositions of the period sought to differentiate themselves from the earlier random displays of "cabinets of curiosity" and instead to show the order and rationality of nature as discovered and studied by empirical science. The theories of Darwin and the corresponding advances in the various sciences that occurred in the second half of the 19th and the early 20th century are intimately bound up with the displays that museums chose to create at that time (Bennett, 2004).

The Impact of Evolutionary Theory on the Display of Human Cultures

Darwin's theories of speciation—of the development of more complex forms out of simpler ones—and of natural selection both reflected earlier theories of human social development and were applied wholesale to the examination and evaluation of human cultures (Shipman, 1994). In the second half of the 19th century the theories of Herbert Spencer, who actually coined the term "survival of the fittest," came to be used to explain differences both among and within cultures. These ideas, known today as Social Darwinism, were used to explain and justify the domination of "primitive" peoples by the "developed" countries of Europe and to explain poverty and wealth in terms of "laws of nature" that should be allowed to play themselves out without government intervention (Shipman, 1994).

At this time many of the large comprehensive museums in Europe, North America, and Australia began reorganizing their collections, not only those of natural specimens, but of human cultures, based on an evolutionary progression (Bennett, 2004). According to this categorization, cultures—particularly those of people of darker skin color—were like fossils in the natural world: while existing in the present, they encapsulated the remnants of early human development and hence were labeled "primitive" or "uncivilized." Other cultures such as those of China or India were judged to be further along on the evolutionary scale, with Western culture viewed as the pinnacle of human development and civilization. The development of the concept of "race" and its conflation with culture also served to bolster these evolutionist views of human development, with nonwhite peoples allocated to the bottom rung of the evolutionary ladder and lighter-skinned peoples the top (Bennett, 2004).

This view of the evolution of human societies has been largely discredited by 20th-century studies of ancient cultures by anthropologists and archeologists and by greater understanding of the biological unity of the human family through research on DNA and the human genome. Yet this approach has persisted in museum displays well into the 20th century, and the residue of these beliefs remains in many of our exhibition halls today with labels such as "primitive art" applied to masks and sculptures that were created in the 20th century but may have been made according to long-held traditions.

TWENTY-FIRST-CENTURY CHALLENGES TO THE IDEA OF A MUSEUM

The Challenge of Culture

> Over the last 20 years a "culture of cultures" has emerged, representing an important frame of reference for communities worldwide: indigenous peoples, ethnic minorities, transnational alliances . . . interest groups. . . . The culture concept is used by groups to fight for recognition, financial support or economic and political rights. (Breidenbach & Zukrigl, 1999)

From the Diorama to the Main Gallery

In the 18th- and 19th-century encyclopedic view of man and nature, the story of Europeans was the stuff of history. Non-Europeans, especially those conquered by Europeans, were assumed to have no history, or at least none before European contact. Their material culture was studied and collected by anthropologists, and came to reside with, and be displayed alongside, the flora and fauna of the natural world (Bennett, 2004). The story of how many peoples have moved from display in natural history museums to display in exhibitions over which they have ownership and control is an important phase in the history of museums.

For example, the Smithsonian Institution, founded in 1846, came to be known as the National Museum in 1881, and consisted primarily of a collection built through westward explorations and expeditions amassing specimens of nature and the material culture of Native American peoples. Over time its collections came to include icons of national pride such as the Star-Spangled Banner and objects that demonstrated the technological advancement of the United States. When in the mid-20th century it was decided to create separate museums for different types of collections, those items of material culture seen as related to American history were moved to the National Museum of History and Technology (later named the National Museum of American History). The material culture of Native Americans was retained in the Department of Anthropology at the National Museum of Natural History (Karp, 1991).

> Thus the Smithsonian, unintentionally but palpably, maintains a nineteenth-century evolutionist distinction between those cultures that are best known and exhibited as part of nature, primarily Native Americans and peoples of the Third World, and Americans whose primary defining feature was originally conceived as the passion of science and technology, and who now possess history, in contrast to natural history. (Karp, in Karp & Levine, 1991, p. 377)

The approaches taken toward displays about indigenous peoples by the Canadian Museum of Civilization and the Australian National Museum at the end of the 20th century reflect this ongoing discussion about who possesses history. The former, after initially planning a hall that showed the entire sweep of Canadian history, decided to create a separate First Peoples

Hall and a Hall of History (later named the Canada Hall.) The latter also has a First Australians Gallery and has made a conscious attempt to intertwine references both to early peoples and to Europeans in all sections of the museum, endorsing a deliberate blurring of the traditionally separate disciplines of anthropology and history (Dean & Rider, 2005).

These accounts place the creation of the National Museum of the American Indian on the Mall in a wider context and make more understandable its interest in telling its own story rather than presenting a story created by Western experts. This, in turn, is part of a larger shift in the ways in which museums hold and display the material culture of non-Western peoples.

> Over the last half century museums have paid attention to these varying sociological theories [multiculturalism, pluralism, "mosaic," "melting pot"] and continuously, through a series of self-conscious adjustments, have tried to reflect the current thinking most in vogue. . . . What followed was a tentative and then more widespread recognition that the descendants of the makers of the objects have a right to share authority in museum display, collections care, and museum management." [These changes have allowed] . . . the voices of minority peoples to become loud enough, and their power to coalesce enough, to cause actual change in museum policies. Museums are learning, in small and sometimes large ways, to share control over their objects. (Gurian, 2004, pp. 89–96)

The Creationist Challenge

Why Has This Become an Issue?

Although the basic theories of evolution began to appear in museums in Europe and the United States by the turn of the 20th century, it appears that they were present primarily as organizing principles for the display of natural specimens as well as human material culture. While such presentations went uncontested for decades, in recent years, particularly in the United States, natural history and science museums, as well as zoos, have struggled with the presentation of evolution. Why has this become such a "hot button" issue?

In the 1960s and 1970s a number of books published by Christian fundamentalist authors John C. Whitcomb and Henry M. Morris brought a resurgence of creationist[1] beliefs to prominence. "The translation of this creationist literature into many languages fostered the rebirth and subsequent spread of creationist beliefs across the industrialized world. Within the United States, creationist and evolutionist beliefs are almost evenly distributed in the population at large, though there are regional variations" (Cavanaugh, 1985, cited in Evans, 2001, p. 219).

Linking Creationism and Science

According to philosopher and cultural critic Bernard Henri Levy, the fuel that has ignited the current debate about creationism and evolutionary theory is intelligent design, which attempts to bring scientific underpinnings to creationism. According to Levy, creationists in earlier periods justified their beliefs by faith, not science. Today, they speak of "creation science" and propose intelligent design as an explanatory system that should be an equal alternative to evolutionary theory (Levy, 2006).

How Is the Creationist Challenge Affecting the Museum Field?

The creationist challenge is affecting the museum field in a variety of ways. An especially important challenge is the argument that it is "only fair" that museums display creation science or intelligent design along with evolution (Scott, 2005). A second challenge is the emergence of creationist museums, many using the latest techniques in museum design, with scientifically correct reproductions of dinosaurs and other early reptiles, placed in settings that also include models of early humans and timelines that date the Earth to 6,000 years (Anderson, 2005; *Chicago Tribune,* 2005; Johnston, 2005).

> Why is the human mind (at least, the Western, protestant mind) so susceptible to creationism and so comparatively resistant to naturalistic explanations for the origin of species?
>
> —E. M. Evans, 2001, p. 252

The most thought-provoking challenge is the use by creationists of evolution, geology, and other natural history exhibitions for their own teaching purposes. This use supports the findings of museum researchers such as

Randi Korn (Jennings, 2005), developmental psychologists such as E. Margaret Evans (2001), and science educators such as Eugenie Scott (2005) that creationist thinking is virtually impervious to information about evolutionary theory.

For one thing, it appears that all young children are "essentialist" in their understanding of the natural world (Evans, 2001). It makes sense that cows, dogs, and people were specially made, just like dolls or cookies. In addition to being compatible with how we all learn and grow developmentally, the belief that the world, and especially human beings, was specially and directly created is consistent with other aspects of Western belief. It is supported by the two major religious traditions that have shaped Western thinking, not just about the natural world but also about the nature of time and history. These religions (Judaism and Christianity) present, both through Genesis and through broader teaching, the idea that the universe has not only a beginning but also a purpose and an end (Evans, 2001). Thus creationism is part of a more comprehensive network of beliefs that was once more widely held but that continues to have deep meaning for large numbers of people.

A Continuum of Religious and Scientific Thinking

In fact, many people in America today live with cultural understandings that take some ideas from the Western scientific tradition and some from Western (or non-Western) faith-based traditions (Evans, 2001; Scott, 2005). For example, people who believe that the Book of Genesis explains literally the origins of the universe do not necessarily reject all scientific explanation for causality. They expect a mechanical explanation for the breakdown of a car, or a physical explanation related to heat or time for why a cake rises or falls. On the other hand, those who believe generally in evolutionary theory and in the validity of the scientific method may find that germ theory, biology, and anatomy can explain why someone is sick or has died. However, these same people may seek nonscientific explanations or ways of coping with the far more devastating question of why their own child has died (Tickle, 2005). Science is not really suited to examining questions of ultimate causality.

A large number of Americans either fall somewhere along an evolution/creationist belief continuum, or believe generally in evolution but are confused or undecided. It is these people that museums need to target (Scott, 2005). Here is where a "cultural" view of creationists may be helpful to museums. Correct information and direct instruction are useful in changing incorrect understandings of simple facts—that is, that two and two equals five, or that "mouses" is the plural of "mouse." However, creationism is not simply an incorrect fact but part of a complex system of beliefs and assumptions—a cultural perspective, if you will—that even multiple visits to evolution exhibitions are unlikely to change. In the presentation of content, science and natural history museums must hold true to their mission of displaying and communicating the best that current scientific research has to offer in all disciplines. At the same time they must appreciate the complexity of this issue. For example, in preparing staff to interact with creationist visitors, museums may find that training in cross-cultural communication and dialogue may be as helpful as having the facts of evolution at their fingertips.

Eugenie Scott, director of the National Center for Science Education and a leader in countering antievolutionists in the public schools, decries a view that places creationists and evolutionists in complete opposition to one another. Instead, she has developed a Creation/Evolution Continuum (Scott, 2005) for teachers, which she recommends for museums also. This continuum recognizes that potential students (or visitors) actually have a broad spectrum of beliefs about creation. On one end are those who hold that their faith requires the acceptance of the flat-Earth beliefs of the authors of Genesis. At the other extreme are "philosophical materialists" who hold that the theory of evolution disproves the existence of God. In between are multiple levels of belief in the Bible and in evolutionary theory, many of which are compatible with one another. Scott's encouragement to classroom teachers to find those areas of compatibility may be excellent advice that museums can follow as well.

IMPLICATIONS FOR MUSEUMS

Museums should explore and articulate the concept of cultural perspective with regard to their history as well as their current theory and practice. Because this concept can provide insight and direction for dealing with some of the most difficult issues that museums face today, museum professional associations should foster research, publication, and conference discussion of this topic. It is important for all museum professionals to have a greater understanding of the very particular values and assumptions out of which museums grew and which are part of the general public's understanding of museums today. It is by unpacking, studying, and then re-articulating these assumptions for the 21st century that museums can begin to address many of the fissures in the larger society that threaten their very foundations.

It is particularly important for science-based institutions to articulate and elucidate that science is a worldview, a perspective, a set of ideas and processes. While it is the mission of science museums to present and promulgate this perspective, they need not and should not give the impression that science is antithetical to religion or that it is the only valid way of understanding the world (Scott, 2005). A perceptive letter to the editor, following an article on the evolution/creationism debate, states:

> Children should not be taught creationism, which is irrational, but neither should they be taught that scientific investigation is our only tool of thought. Imagination can also bring us close to the unknowable powers of the universe, of which we are a part. (*Manchester Guardian*, 2005)

Museums' stances toward their visitors should be nuanced and empathetic to the varying points of view brought to the understanding of their subject matter, for example, the theory of evolution. In the tradition of constructivist learning that is considered best practice in museums today, those who develop exhibitions, materials, and programs must consider the differing perspectives that visitors bring. Age, education, and the influence of family and community shape visitors' levels of prior understanding and readiness to grasp a topic such as evolutionary theory. The use of front-end evaluation as part of the exhibition development process, the convening of advisory groups from various constituencies, the regular assessment of programs and materials developed for visitors—all of these can provide museums with a deeper understanding of how best to communicate with the communities they serve. It is the museum's responsibility to present fully what is defined by its mission and at the same time to understand that not all will see the world as the museum displays it.

Museum training for new professionals as well as current staff and volunteers should include training on the concept of cultural perspective and ethnocentrism. Whether the staff is dealing with creationist families who want to use natural history displays for their own teaching, with people who are encountering the traditional cultures of Native Americans for the first time, or with non-American visitors who are curious about why Americans seem to place such emphasis on their flag, having an understanding of the concept of cultural perspective is essential. This approach encourages the presenter to understand that the way he or she views the world is *a* viewpoint rather than *the* viewpoint. It gives permission to allow for multiple points of view without denigrating one's own. It encourages the framing of information in an interpretive rather than prescriptive manner.

Museums should carefully define or redefine their missions and articulate them prominently to the public. At a recent Association of Science-Technology Centers (ASTC) conference meeting (Jennings, 2005) on this topic, museum directors Emlyn Koster and Eric Jolly, museum researcher Randi Korn, and museum consultant Elaine Heumann Gurian each called for museums to define their missions and then make decisions on what to include or exclude based on those missions. Randi Korn (2005) summarized these thoughts in what she calls "the Intentional Museum."

> The Intentional Museum has an institution-wide accepted mission; lives by a philosophy and principles that guide its decision making regarding collecting, content, interpretation, and community relationships; and acts decisively when issues emerge. An Intentional Museum knows how it will respond to its publics—no matter the issue; articulates how it wants to be known and experienced by others; describes the types of experience it wants to create; has structured its management strategies and practices so visitors realize these experiences through exhibitions and programs; and continually assesses the relationship among institutional intentions, institutional practices, and visitor experience outcomes. It goes without saying that an Intentional Museum has a leader who works with staff to negotiate the museum's intensions and creates an open work environment where debate and risk taking are welcome activities.

ANOTHER LOOK AT OUR OPENING QUESTIONS

1. *In the face of opposition from creationist visitors, how can science and natural history museums remain true to their institutional missions? Should they try to serve all of their constituents, including those who reject important elements of modern science?*

Science and natural history museums must stand firm in holding to their mission to present the most current findings in science. They can also present an empathetic and welcoming face to all of their visitors, train their staff to work respectfully with people of diverse viewpoints, and seek out those areas where science and religion find common ground.

2. *Why is it that, increasingly, the creation narratives of non-Western communities are given respectful display in major museums while the museum field is generally resistant to the requests for "equal time" by proponents of creationism or intelligent design? Is this a principled stance or a hypocritical position?*

Science and natural history museums should not feel obligated to give equal treatment to creationism in any of its forms, including intelligent design. Creationism in its varying configurations is ultimately based on faith, a mode of understanding that is outside the purview of science. Presenting such views within a science or natural history museum as *alternate scientific theories* would give them an epistemological basis that they do not possess. On the other hand, exhibiting other cultural or religious views of creation and the natural world *on their own terms and not as science-based theories* is certainly a legitimate role for museums. As this chapter has shown, the presentation of multiple cultural perspectives on their own terms and in their own voices is an increasingly important role for museums in our global culture.

One might imagine an exhibition in a cultural museum or a natural history museum that displays a variety of creation stories from around the world, including the biblical account. The creation story in Genesis, for example, could be presented in terms of the tremendous impact it has had on worldwide art and literature, not to mention its role in shaping a view of human beings as having "dominion" over the natural world. Such an exhibition would no doubt raise objections from those who might resent the biblical story being placed on an equal footing with other creation accounts. But such a presentation would locate the creation story in its proper cultural context and take it out of the realm of empirical knowledge, where it does not belong.

3. *What do these two questions say about the nature of the museum itself? To what extent can and should museums remain a kind of embodiment of the Western scientific tradition? And in what cases should these old skins expand to contain new wine?*

Reflection on these two questions should give one hope for the museum field as a whole. Museums need not and should not remain tied to one cultural tradition. Instead, people with many varying points of view in communities all over the world continue to create museums as expressions of their values, identities, and ways of knowing. This is a positive direction, as long as new museums align their modes of display with their missions and goals. The centuries-old "skin" of museums seems infinitely capable of containing new vintages from the global vineyard.

ENDNOTES

Special thanks to Randi Korn and Elaine Heumann Gurian for raising the questions on which this chapter is based and for their advice, editing, and encouragement throughout. Thanks also to Carol Stepp for her interest and assistance, especially in the early stages of this paper.

1. While there is a range of ideas that can be called "creationist," I am using the term to mean belief in the special, direct design of humans and other species by God. Some believe that this happened in six days, others over millennia, but the common thread is the belief in direct design. Intelligent design is a form of creationism that accepts many aspects of evolutionary theory but maintains that evolution is insufficient to explain the complexity of nature, which proves that there is an intelligent designer (Scott, 2005).

8

Raising the Relevancy Bar at Aquariums and Science Centers

Emlyn H. Koster and Jerry R. Schubel

MAXIMIZING RELEVANCE: CONTEXT AND AIMS

Among the most iconic images of the modern scope of science and technology, and by implication of society and the environment, are the spectacular photographs of Planet Earth by Apollo astronauts during their 1969–1972 missions to the moon. In 1948, Fred Hoyle, the distinguished British astronomer, predicted that once a photograph of the Earth from a distance became available, a new line of thinking as powerful as any in history would be let loose. NASA's images of Planet Earth indeed spawned a revolution of philosophy and action on global matters, in particular the responsibility of protecting natural environments driven by the obvious fragility of our blue-green world in space with its wafer-thin atmospheric envelope.

The fact that *Scientific American* used one of these early NASA images on the cover of its September 2005 special issue on the theme of Crossroads for Planet Earth was confirmation not only of their enduring nature but also that they serve as a poignant reminder that we have yet to learn as much as we must about global stewardship and, perhaps even, our very survival. The opening editorial in this issue offered this provocative view (*Scientific American*, 2005):

> Geographer Jared Diamond's recent book *Collapse* documents past civilizations that could not recognize or bring themselves to change unsustainable

> ways. Largely because of science, our civilization has the chance not only to avoid their fate but to enter an age of unprecedented prosperity. Science is not and should not be the sole factor in decision making; others, such as moral values, are also crucial. But we need to go into these decisions with our eyes open to what is going on in the world.

The almost four decades since that Apollo mission have been marked by an immense rate of scientific and technological advances, including instant communication, steadily increasing rates of disruption to natural environments, mounting concerns that human activity itself is exerting a global climatic impact that can be catastrophic, initial planning steps for a world that many energy experts say has already surpassed peak production of its fossil fuel supplies, dwindling seafood supplies, a widening gulf between affluent and poor societies, and escalating tensions among the world's different religious groups.

Countless eminent scientists around the world have shared their concerns about the widening gap between society and science and the prevalence of short-term thinking in public policy circles. For example, Ursula Franklin, a Canadian, has eloquently stated (1990), "The task of the future is to build knowledge and understanding among and between citizens and scientists, so that the distinction between the two groups disappears—so that both become citizen scientists, potentially able to solve our problems together."

And so, with this introduction, we turn to the role of aquariums and science centers with a theme of raising their relevancy bar. As one of us has recently pointed out on the occasion of the centenary of the American Association of Museums (Koster, 2006):

> Each museum has a choice of overall direction and external contexts from an array of possibilities. It is clear that the world—on local, regional and global scales—has myriad opportunities for improvement as well as challenges to try to overcome. That these correspond to the subject areas of the different types of museums that have evolved over past centuries and decades is also clear. Fundamentally then, each museum has a choice of the degree to which it will increase its external orientation to address contemporary and future matters, both locally and globally. Museums that become more relevant are likely to attract more robust funding and therefore become more sustainable and more valued institutions. There are factors at

play that both encourage and discourage this stance. Greater currency of content raises the possibility of controversy, and controversy could trigger a mood of retreat. But another, arguably more attractive perspective is that the museum profession can make an increasingly bold and supportable contribution in helping society to understand and improve our collective future through different approaches and extensive collaboration.

We start with a working definition of "relevancy." For our purposes, relevance is sensibly defined as the logical connection that one thing has with another, as, for example, the orientation of an institution's philosophy and practice to prevailing issues—locally, regionally, and/or globally. It follows that an institution is not properly justified in using this descriptor unless it is comfortable and capable of tackling contemporary and consequential subject matter (Koster, 2006). In this chapter, we explore the connections of our institutions with real-world issues, current events, and philosophical trends.

Aquariums and science centers are mostly structured as nonprofit institutions with their success measured primarily in terms of mission advance. Financial viability, however, is a prerequisite to mission success, as the often quoted maxim of "no margin, no mission" reminds us. Since most of our institutions rely extensively on visitor admissions for financial health, it is clear that we must be viewed by our visitors as having relevance to their lives—to their needs, wants, and desires. In today's world, with its widespread talk about the "dumbing down" of expectations, the role of our institutions to try to elevate the intellectual level of discourse becomes an important consideration. Recently in the United States, both the National Science Board (2005) and the National Academies (2005) have issued benchmark reports to this effect. It is now widely in the mainline news that the United States has fallen behind in science education in schools and in adult science literacy (Lemonick, 2006). In 2000, America's first astronaut John Glenn was commissioned by the U.S. Secretary of Education to chair the National Commission on Mathematics and Science Teaching for the 21st Century. The executive summary started with this stark perspective (National Commission on Mathematics and Science Teaching for the 21st Century, 2000): "In an age driven by the relentless necessity of scientific and technological advance, the current preparation that students in the United States receive in mathematics and science is, in a word, unacceptable."

We must also remind ourselves that relevance is both in the eye of the "receiver"—the visitor—as well as in the minds of those in leadership positions who are accountable for doing the right thing in a big-picture, long-term context. The pursuit of relevance must be based on an understanding of the communities that the science center or aquarium seeks to serve and what the needs and wants are of those communities on different temporal scales in the context of the vision and mission of the organization. While there should be consultation with users, relevance often cannot be determined solely by focus groups and customer surveys. Certainly, the pursuit of mission-advancing goals that are lofty but that do not succeed in drawing adequate audiences is not fiscally sustainable. Since most of our institutions aspire to repeat visitation, this perspective on relevance takes on added importance.

Surely the optimal key to the meaningful success of our types of institutions is an ability to nurture a balance between the dynamics of popularity and usefulness.

THE RELEVANCY AGENDA FOR AQUARIUMS

The U.S. Commission on Ocean Policy (2004) and the Pew Oceans Commission (2003) both made a compelling case that the condition of the world's oceans, particularly the coastal oceans and the Great Lakes, are in decline and that any sustained solutions must be rooted in an informed, concerned, and committed public. Both reports pointed to (1) the woeful state of awareness and understanding of the public on how the ocean affects them and how they affect the ocean no matter where they live, and (2) the importance of aquariums in enhancing the ocean literacy of the general public, in addition to the K–12 population. The U.S. Ocean Action Plan is the Bush Administration's response to the U.S. Commission's report. It also calls for efforts to enhance ocean literacy (2004). The relevance of nonprofit, mission-driven aquariums committed to conservation must be measured, in part, by the extent to which they are engaged in making the recommendations of these reports a reality.

In June 2005, the Consortium for Oceanographic Research and Education (CORE) and the Aquarium of the Pacific launched an initiative to raise the level of ocean literacy among residents of Southern California from Santa Barbara to the U.S.-Mexico border. This effort built on the "Essential Principles and Fundamental Concepts" developed by the Centers for Ocean Science Education Excellence (COSEE) and the National Marine Educators Association (NMEA). CORE and the aquarium assembled 40 marine scientists, most from Southern California, to identify and describe what they believed every resident of the region should know about how the ocean affects them and how they affect the ocean. Not surprisingly, their documentation was comprehensive and extensive and in far greater scientific complexity than our institutions could reasonably present to the general public (Schubel, Monroe, & Lau, 2005).

This led to a second workshop that brought together experts in informal science education modalities (exhibits, programs, field experiences, the media). Their challenge was to translate the output of the scientists into storylines that would engage, educate, and empower the general public. The participants from aquariums, science centers, wildlife refuges, and other conservation organizations conceptualized how to make the content from the first workshop relevant to a visitor. They said this about telling ocean stories that will result in stewardship:

> Stories about our ocean are more "complete" than simple facts. Stories provide a context in which the public can connect new knowledge to what they already know. Journalists write "news" stories, advertisers tell "product" stories. Biographers tell "life" stories; novelists write about intrigue, mystery, romance, etc. Explorers tell stories about their adventures. What stories do informal ocean science educators tell?
>
> If our stories are to compete with the many others that are presented to the public each day, ours must grab and hold the attention, be fascinating as they unfold, and contain a take-home message about needed stewardship from an "in my backyard" and regional focus. They must also be viewed as coming from a credible source. Hopefully, when the story has ended, the visitor who shared in its telling will not only want to tell it to others, but also want to take the steps necessary to "walk the talk," i.e., become a steward for the ocean. (Monroe, Lau, Schubel, & Cassano, 2006)

Aquariums do not have a strong history of working together as a network, but there are encouraging signs. A coalition of four of California's five major aquariums recently formed the California Aquarium Collaborative to promote public ocean literacy. The partnership is committed to developing and delivering a set of common messages in combination with regional messages that should have particular appeal to their visitors. And Coastal America's Coastal Ecosystem Learning Centers (CELCs), COSEE, and Sea Grant have joined together to host a series of regional meetings patterned after the Southern California model. By working as networks of free-choice learning institutions, we can multiply the power, reach, and relevancy of our programs.

On June 7–8, 2006, a national *Conference on Ocean Literacy* (CoOL) took place in Washington, D.C. The conference was coordinated by the National Marine Sanctuary Foundation in partnership with the White House Council on Environmental Quality and the White House Office of Science and Technology Policy, and had broad sponsorship by federal agencies. While the conference report will not be available for several months, one of us (Schubel) was on the steering committee and a participant on the informal education panel. The following comments are based upon recommendations at the conference that are particularly relevant to aquariums (and other informal science organizations) in promoting ocean literacy. Aquariums should focus on:

- raising awareness—a prerequisite to learning
- learning and not on teaching
- creating experiences that involve heads, hearts, and hands
- instilling a sense of awe and wonder for the ocean and ocean life
- giving visitors outdoor experiences to introduce them to nature first-hand

They should also:

- partner not only with other aquariums, but with different kinds of organizations, and with the media, and
- provide context and perspective for ocean issues utilizing local and regional "hooks."

If our institutions are to be relevant, another area that demands attention is the loss of the competitive edge of the United States in the fields of science, technology, engineering, and mathematics (STEM) fields. The key to reversing the current trend must come through capturing the imaginations of young people by middle-school years about careers in these fields and by providing them with educational opportunities at critical times. Aquariums have the capacity to play a much larger role in the nation's educational enterprise than we do. We should be able to bring the excitement and the magic of STEM fields alive for young people. We should form stronger collaborations with leading scientists, engineers, and mathematicians not only in universities but also in business, industry, government, and non-governmental organizations (NGOs). Many of these career trajectories may have greater appeal to young people than careers in academia.

For each of the past three years, the Aquarium of the Pacific has had an intensive ten-week program for fourth and fifth graders in which they design a "sustainable Long Beach" for the year 2025. Kids have access to city planners, the staffs of city departments, hospitals, the Port of Long Beach, and others who act as their consultants. They elect their own mayor, commissioners, and other officials. In the final session they present their model city and plan to the current leaders of Long Beach. The model and the plan are the outputs. The desired outcome is a future generation of citizens committed to living in harmony with their environment in a sustainable community.

Aquariums and science centers make the "real" world relevant to their visitors by taking them outside and into the natural and built worlds. It is particularly important in urban areas to make people aware that nature exists—maybe not wilderness, but nature, which is everywhere. Outdoor experiences provide a milieu that nurtures interactions among people of diverse ages, experiences, socioeconomic strata, and cultures. The wide variety of outdoor locations and potential activities allows a multisensory intimacy with rivers, beaches, coastal environments, and the ocean. The chance to see, hear, smell, taste, and touch nature is important, as has been demonstrated in the recent book by Richard Louv (2005).

Because outdoor ocean-related experiences engage people, they facilitate learning about the ocean, the problems, and the role of the individual in solving those problems. By stimulating curiosity and the desire to explore the natural world, outdoor experiences nurture a keener sense of observation and greater awareness. Aquariums provide a number of outdoor ocean-learning experiences such as whale-watching trips, birding by kayak, beach cleanups, wetland restoration projects, and other outdoor experiences that strengthen personal connections with the environment and promote stewardship.

Like science centers, aquariums provide the public with opportunities to venture far from home on ecotours to exotic places. Staffed by scientist lecturers and aquarium naturalists, these trips introduce participants to a different natural world and help them understand their relevance to the global ecosystem. The combination of these trips with trips to local environments is powerful. For those individuals who cannot venture far afield, many aquariums bring in scientists, journalists, photographers, explorers, and authors to share their work, knowledge, images, and concerns about the world's oceans through lectures, films, and panel discussions on a variety of subjects.

Long Beach is one of the most ethnically diverse large cities in one of the most ethnically diverse regions of the country. Four years ago we looked at the demographics of our visitors and found they did not match the community we seek to serve in terms of ethnic diversity, economic diversity, and educational diversity. We set out to change that. We consulted with people who came, and those who did not, to find out what would make us more accessible, more interesting, and more relevant. In response, we created a series of weekend cultural festivals to celebrate the diversity of our city and region through music, dance, storytelling, and food. We created new signage and guidebooks in Spanish. We also created new discount programs for those unable to afford the normal admission. We now have one of the two most ethnically diverse aquarium audiences in the country.

THE RELEVANCY AGENDA FOR SCIENCE CENTERS

The term *science center* was first used in Seattle following its 1962 international exposition. The science-themed American pavilion was so popular that city officials chose to continue that experience after the exposition, and the Pacific Science Center was born. Canada's centennial project in Toronto of a museum about science and technology became the Ontario Science Centre. It opened in 1969, the same year as the Exploratorium in San Francisco.

Starting with these institutions, there has been a veritable explosion of science centers. The Association of Science-Technology Centers began in 1973 and today lists 500+ member institutions in 40 countries. About a dozen regional science center associations have also been formed to enable more culturally specific networking at more accessible conferences. Additionally since 1996, triennial Science Centre World Congresses have taken place in Helsinki, Calcutta, Canberra, Rio de Janeiro, and, next in 2008, in Toronto.

This impressive growth tends to suggest that science centers are doing very well. Certainly they are popular institutions, but the question must be asked if they are also being maximally useful. It is instructive to recap how the philosophy and practice of the science center field have evolved.

The first participatory approaches in science centers, which predated computers, were to enable visitors to understand basic scientific principles. Mechanically interactive exhibits about physical forces, electricity, and magnetism were commonplace. Because of the centrality of these principles to schools' science curricula, teachers were attracted to bring class field trips to science centers in large numbers. Families with young children followed suit. To learn by individually experiencing these phenomena was a powerful alternative to textbooks and laboratory instructions, and it was enjoyable.

The adult generation of the day was accustomed to traditional science museum experiences of specimens and artifacts in display with text panels, and the occasional demonstration that they could watch. For many adults, interactivity was not an inherently comfortable expectation of their mode of visit and as exhibits began to include computer-based stations, science centers became increasingly identified with a young audience in the general public's mind. One of the most popular images of a science center experience became the smiling young child with her or his hair projecting outwards while attached to a static electricity generator. Indeed, more than three decades after the birth of the science center movement, this image persists in many promotional materials.

Not surprisingly, it became common for science centers to use the word "fun" in their marketing taglines to attract new generations of child-centered audiences to their engaging experiences. One can also readily see in hindsight that a temptation to blur the boundary with entertainment was bound to occur. Simulators with thrill ride and science fiction experiences began to appear, giant-screen theaters started to include films unrelated to the institution's science educational mission, and sensation started to become acceptable in the subject matter of traveling exhibitions. Science centers then started to adopt a variety of positions to actively justify these kinds of experiences with statements such as "Science fiction attracts visitors not normally attracted to a science center," "Visitors can distinguish between the learning and entertainment dimensions of the science center experience," and "The institution needs the extra box office revenue to survive."

The unfortunate irony of this evolved orientation of the science center field is that it has been during the very decades of their development that the volume of science-based challenges and opportunities in the world has risen exponentially. In hindsight at least, the disconnect between the content of science centers and the at-large situation facing the world has become alarming. It would seem that the popularity of interactivity, which has been the hallmark of science centers, as well as a reluctance to be controversial across the museum field, have both lulled science centers into seldom wishing to confront science and society subjects of profound consequence. These include issues of lifestyle choice, the widening gulf in access to life-enhancing technologies between poor and rich regions of the world, the finiteness of fossil fuels and the necessity of viable alternatives, human-caused collapse of most natural ecosystems and the accelerated rise in sea level, the direction of human evolution, and the psychology of bias, racism, and terrorism.

As with aquariums, to be maximally useful, science centers should choose content that is meaningful to living, learning, and working in the surrounding region and ensure that it is accessible to the full cross-section of the surrounding communities. In many ways, science centers have been able to accelerate the relevancy trend in museums because they do not have defining collections, they exist entirely for a public purpose, and they can readily use technology to amplify and extend their learning experiences (Koster, 1999).

After a $68 million capital campaign, Liberty Science Center in New Jersey opened in 1993 to great fanfare, instantly becoming the state's most visited museum and soon becoming a recommended destination in guides to New York City. But then financial alarm bells rang and a structural deficit was in urgent need of fixing. Three years after opening, a review of zip code attendance data from counties in a 100-mile radius revealed an unacceptable underrepresentation of students from Jersey City, the host community, itself a designated at-risk school district. Annually, less than 2% of the city's public school enrollment of 32,000 used Liberty Science Center. As a result of a proactive statewide partnership program, the figure increased to 60% of enrollment. Under the terms of state-funded, whole-year service agreements, the center provides onsite, offsite, and online resources to participating schools as well as teacher professional development and educational opportunities for the families of all district students.

What started as the probing of attendance data became a transformation of consciousness within Liberty Science Center. Step by step, this institution grew a portfolio of socially responsible programs geared to several pressing regional needs (Koster & Baumann, 2005). In so doing, the institution also naturally developed a diverse audience representative of the New Jersey and New York City metropolitan region. When the terrorist attacks occurred on the World Trade Center a mile away across the lower Hudson River, a response to help in all ways possible over the ensuing days, weeks, and months, until New Jersey's Family Assistance Center at the riverfront ended its services, seemed automatically to be what Liberty Science Center should do as an adaptive public service priority (Koster, 2002).

At the center's 1989 groundbreaking, then-Governor Thomas H. Kean stated to the news media that he anticipated a doubling of Liberty Science Center's size within a decade to meet a growing educational demand for its learning and teaching resources. Twelve years after its 1993 opening, a $104 million facility expansion, exhibition renewal, and program enhancement project got underway. Seven years in planning, this shared public and private sector project is seen by the New Jersey state government as tackling pressing needs in science education, workforce development, public literacy, and the quality of life. That a private, nonprofit corporation has received major capital funding from state government while also continuing to receive annual contractual funds from it for systemic service to all designated at-risk districts would seem to confirm that the relevance and sustainability of a science center are indeed intertwined.

A new suite of exhibitions as well as new educational facilities will occupy the completely refreshed 295,000-square-foot building that is scheduled to reopen in July 2007. The Center for Science Learning and Teaching is intended to reshape how science education is perceived in the region. Unlike anything available at schools, its new facilities will be a gateway where students and teachers will be exposed to scientists and science content in a manner that will stimulate learning and career choices.

There will also be a theater where students have videoconferences with operating room teams at nearby hospitals as cardiac bypass, kidney transplant, and neurosurgery takes place.

Each new exhibition theme has been chosen to be relevant to living, learning, and working in the surrounding trans-Hudson region, one that stretches from Albany, New York, to Hartford, Connecticut, to Harrisburg and to Philadelphia, Pennsylvania. *Our Hudson Home* is about how stewardship of the local river and estuary requires understanding of their complex habitats, and of the human and commercial activities they support. *Skyscraper! Achievement and Impact* explores the design and engineering of tall buildings and their environment, and how these are evolving to better address human needs. *Infection Connection* is about how individual and collective choices determine the impact of infectious diseases on the health of people around the world. *Communication* explores how continual development and applied ingenuity help us learn from the world around us and create ways of communicating with one another. *Breakthroughs* profiles how the convergence of science, technology, and society necessitates the public's understanding of new discoveries and milestones and the impact these have on people's lives. With an assortment of live animals, *Eat and Be Eaten* explores how species evolve appearances and behaviors that help them avoid detection from predators and as prey. *I Explore*, for preschoolers and their caregivers, exposes young minds to the scientific method and to science as a human endeavor, as well as enabling them to discover who they are, where they are, and what they can do. In a new *Exhibit Commons* concept, visitors and others anywhere else around the world will be invited to submit ideas for altering and adding content to the current exhibitions.

In terms of a contribution to raising the bar on relevance, Liberty Science Center is indeed fortunate to have the opportunity of a large-scale total renewal. However, it is fair to add that this project is, in essence, a direct outcome of Liberty Science Center's journey over the past decade to expressly focus on initiatives rooted in a principle and a practice of external value.

A SUMMARY VIEW

This discourse about raising the relevancy bar in aquariums and science centers is symptomatic of a growing atmosphere of unease across the museum field (Weil, 2002). Whether prompted by a conscience over the troubled state of society or the deterioration of environmental conditions, or whether prompted by funding sources or a financial pressure that encourage a fresh outlook, museums have considerable untapped potential to play more important roles as cultural institutions. For their part, as examples, aquariums, natural history museums, science centers, and zoos have the capacity to become more sophisticated in both substance and brand terms by being seen to be, and wholeheartedly being, society's primary lifelong resource for engagement with pressing science-driven and environment-driven opportunities and challenges.

Initially, public opinion around almost any new issue confronting society and/or the environment is statistically distributed as a bell curve. Opinions lead to discourse and discourse leads to new perspectives. Perspectives then lead to a broader understanding. Gradually a majority forms, more enlightened decisions are made, and resulting actions are taken. Since the initial issue rarely goes away, society's choice is either one of reasoned confrontation or denial.

To tackle contemporary issues in the museum is certainly to embark on more difficult subject matter and requires new ways of working (Davis, Gurian, & Koster, 2003). An extra measure of innovative effort is required to plan successfully and forge progress. Given the pressing issues that are germane to the missions of each museum type, we contend that it is a matter of accountability whether or not our institutions opt to be part of the solution or part of the problem. In our view, the search for greater relevance in our institutions must never cease.

Successfully heading in this direction depends on three facets of institutional culture being in place. The first concerns mission and vision—is there a clear and firm commitment to be of value to the societal and environmental problems we face? The second concerns leadership—is there a preparedness and competence to be an activist? The third concerns strategy—is there a relentless pursuit to be more externally useful and to nurture new perspectives in funding stakeholders?

At the centennial conference of the American Association of Museums, the noted Harvard Business School professor Michael Porter delivered a keynote address on strategy in the context of museums. He urged that our institutions be clear about who they are, what makes them different, and why and how they exist as part of a value chain in society. In business marketing circles, a value proposition asserts and positions the tangible and differentiated value of an offering. Thinking in this way calls for museums to be of distinctive external value in two integrated respects—namely, what the surrounding region importantly needs that our mission and expertise can provide, and what the region's population cannot obtain elsewhere easily, as a comparable out-of-home social and aesthetic experience, or at all. Surely this is the stance that will maximize the museum's relevance, and therefore its sustainability as well.

9

Challenging Convention and Communicating Controversy: Learning Through Issues-Based Museum Exhibitions

Erminia Pedretti

Recently, museums and science centers have created issues-based exhibitions as a way of communicating controversial subject matter to the public. Examples of installations that have addressed *and* created controversy include: *Birth and Breeding* at the London Wellcome Institute; *Of the People—The African American Experience* at the African American Museum of Detroit; *Dialogue in the Dark*, Consens Exhibitions Ltd, Hamburg, Germany; *Darkened Waters: Profile of an Oil Spill*, Pratt Museum in Homer, Alaska; *Endings: An Exhibit About Death and Loss* at the Boston Children's Museum; *A Question of Truth* at the Ontario Science Centre; and *Mine Games* at Vancouver's Science World (see Pedretti, 2002, for a thorough review and critique of some of these exhibitions, and McConnell and Hess, 1998, *A Controversy Timeline*). Issues-based exhibitions like these challenge conventional phenomenon-based or object displays, creating different learning and meaning-making experiences for visitors.

For the past ten years I have been exploring issues-based exhibitions as a way of investigating the communication of science and technology to the public. A number of studies employing various mixed methodologies were conducted in order to understand the nature of these installations, and consequently, how visitors experience them. For the purposes of this chapter, I revisit my experience and research[1] with two particular exhibitions—*Mine Games* (which opened in 1994 at Science World, Vancouver) and *A Question of Truth* (which opened in 1996 at Ontario Science Centre, Toronto). Specifically, I use science and science centers as a "window" into understanding knowledge formation among museum visitors. I investigate how these installations challenge conventional exhibitions in museums, while developing effective learning environments for teaching and learning about complex issues.

CRITICAL CONVERSATIONS AND NEW DIRECTIONS IN MUSEUMS

A review of the literature (see Pedretti, 2002, for a comprehensive review) suggests that during the past decade museums have focused new attention and resources on developing exhibitions that move in novel and rather provocative directions. Issues-based exhibitions represent a paradigm shift from the "objects in glass cases" to an emphasis on involvement, activity, and ideas (Beetlestone, Johnson, Quin, & White, 1998). In the following section, I describe changes that are occurring in places like science centers and science museums, as a way of arguing for issues-based exhibitions in informal settings.

A new generation of science museums and science centers seems to be emerging (Janousek, 2000; Koster, 1999). Science centers are beginning to see themselves as important players in a number of external scientific, social, cultural, and political contexts: "The science center's role is to seek tools to draw the cultural framework, animate the debate, and promote healthy skepticism over superstition and irrational thinking" (Beetlestone et al., 1998, p. 21). Accordingly, discussion and exhibition development in science museums is evolving to include social responsibility and the raising of social consciousness. For example, in November 1996, the London Science Museum mounted a three-day conference, *Here and Now, Contemporary Science in Museums and Science Centres*, to address the challenges faced by these institutions in presenting contemporary science and technology issues (see Beetlestone et al., 1998; Cannon-Brookes, 1997; Farmelo & Carding, 1997). The conference delegates defined contemporary science as "the science that appears in the mass-media spotlight and so, however briefly, touches peoples' lives: genetic screening, life on Mars, radiation and contamination" (Beetlestone et al., 1998, p. 21). Science then, is framed by culture—its context has social, economic, political, and historical dimensions.

Transforming this kind of content into accessible experiences for visitors requires innovation and vision beyond usual phenomenon-based exhibitions. Typically, there are two types of exhibits (not to be considered as mutually exclusive) found at the science center: experiential and pedagogical (Wellington, 1998). The *experiential* exhibit allows the visitor to *experience* (and perhaps interact with) phenomenon (e.g., soap bubbles, whirlwinds, water vortices, air or water phenomena), while the *pedagogical* category actually sets out to *teach something* (e.g., position of organs in the body, digestive system, or reflection of light). These categorizations are useful because they help map out the terrain of possible exhibitions, and also reflect the dominant traditional way of representing subject matter. Conventional installations, for example, often convey science as void of any social cultural context, and negate raising questions about the status of scientific knowledge. These shortcomings lead to what I would identify as the emergence of a third category—*critical* exhibitions. By that I mean exhibitions that critically delve into controversial subject matter to explore, for example, relationships among science, technology, society, and environment (STSE) (see Cannon-Brookes, 1997; Macdonald, 1998; Mintz, 1995; Pedretti, 1999). Many of these critical exhibitions are issues based, inviting visitors to consider material from a variety of perspectives, and to engage in decision making and healthy debate.

LEARNING IN THE MUSEUM

Critical issues-based exhibitions are often emotionally and politically charged, and call upon a different kind of intellectual and emotional response from the visitor. They reflect a radical departure from more traditional interactive hands-on exhibitions and their preoccupation with immediate sensory experience or explication of phenomenon. A number of theoretical perspectives (see, for example, Csikszentmihalyi & Hermanson, 1995; Macdonald, 1998) are helpful in understanding the learning potential of critical exhibitions. In this chapter, I draw on Falk and Dierking's (2000) Contextual Model of Learning because it elegantly incorporates much of what we know about learning. The model's simultaneous attention to multiple contexts, breadth, and depth position it as a useful framework for exploring the nature of learning, particularly in nonschool settings.

Falk and Dierking's (2000) Contextual Model of Learning involves three overlapping contexts: personal, sociocultural, and physical. Visitors are understood as being actively engaged in the construction and reconstruction of these interweaving contexts. This model emphasizes that learning is situated within a rich context. Learning is not an abstract experience that takes place in a sterilized environment; rather, it is "an organic, integrated experience that takes place in the real world" (Falk & Dierking, 2000, p. 10). Intentionally, the model is more descriptive than predictive. For the purposes of this chapter, I focus on the "personal context" of learning and associated themes that emerge from my research, specifically: personalizing subject matter, evoking emotion, stimulating dialogue and debate, and promoting reflection. Although I am attending to only a small part of the model, it should be emphasized that these overlapping contexts operate in synergy, and it is impossible to treat each context entirely in isolation.

THE CASE STUDY EXHIBITIONS

Two exhibitions that provide a useful context for this discussion are *Mine Games* and *A Question of Truth.* Each addresses contemporary issues in very different ways. In *Mine Games* visitors participate in a simulation and are faced with the challenge of deciding whether a silver mine should be built in a fictional town, and if so, how can the safest, most economical, and most environmentally acceptable mine possible be designed. The most striking feature of this exhibit is *Hot Seat. Hot Seat* is built much like a scaled-down version of an amphitheater. Visitors are seated in a tiered semicircle, facing a mediator who has control over a range of interactive multimedia resources. Ideally, information gathered from the smaller galleries (*Wild Things, High Stakes, Blast-It*, and *Boulderdash*) and questions emerging from the visitors' experiences are discussed as they attempt to reach consensus about whether to build the mine.

The second installation, *A Question of Truth*, can be characterized as having a strong nature of science (NOS) orientation as it explores socio-scientific issues related to bias in the history and practice of science, and the generation of knowledge. The exhibition, with its provocative title, is designed to examine several questions about the nature of science, how ideas are formed, and how cultural and political conditions affect the actions of individual scientists. *A Question of Truth* has created some controversy precisely because it questions the pursuits of science, and acknowledges in a rather confrontational way a historical legacy of moral, ethical, and social repercussions.

LEARNING ABOUT LEARNING THROUGH ISSUES-BASED EXHIBITIONS

The exhibitions *Mine Games* and *A Question of Truth* contribute to visitors' understanding of science and society by considering science and social responsibility, controversy and debate, decision making and ethics. However, this begs the question of *how* and *why* these installations enhance learning. Utilizing the personal context (Falk, 2001b; Falk & Dierking, 2000), I examine four factors that are particularly significant to issues-based installations and consequently to visitor learning.

1. Personalizing Subject Matter

For many, science is perceived (and experienced) as an abstract, linear, and impersonal subject, void of meaning or context. This characterization of science is often replicated in many experiential and pedagogical exhibitions. However, critical issues-based exhibitions, by their very nature, personalize and humanize science.

It has been argued that personalization of science offers a more coherent and meaningful approach to teaching and learning science. Pedretti and Hodson (1995, p. 465) posit that personalization of learning means ensuring that "(i) learning is rooted in the personal experiences of individual learners; (ii) science is seen as more person oriented; and (iii) science education is infused with sound human and environmental values." John, an eighth-grade teacher, echoes these sentiments as he describes the importance of drawing connections and personalizing science for his students:

> We need to make the connection between the technical aspect of science and our general and everyday living. I like to see more linkage between the pure

> science and the applications science. . . . I find that quite a lot of students have a very abstract view of science. Science is something that you do in a lab period and they don't see interconnections between what's happening to them in everyday life and what scientists are doing in the lab. (Interview)

Mine Games and the creation of Grizzly (the fictional town) capture the imagination of visitors and bring to the forefront issues that are very real to people's lives. In British Columbia, the mining industry provides many families with their livelihood, while at the same time, ignites controversy as environmental groups question the long-term effects of mines. John continues his analysis with the following:

> I saw the opportunity with *Mine Games* to be able to make linkages between the classroom textbook materials the kids are receiving, with something a little more practical and something that they can actually relate to, that is happening in British Columbia itself. Okay it's a hypothetical town and a hypothetical situation but a lot of these hypothetical situations are issues that are actually occurring . . . coming out in newspapers every day. Talk about reclamation, talk about huge chunks of land being set aside as national parks, or federal parks or provincial parks . . . the clear-cutting. . . . There's a lot of opportunity for the kids to see and think about real-life problems and how to go about solving them.

Similarly, *A Question of Truth* capitalizes on personalizing science by reaching people through issues such as science, race, and prejudice—issues that spark strong personal connections and reactions, particularly in a multicultural city like Toronto. These are topics that people feel passionate about, and therefore levels of engagement potentially increase (Pedretti, 2002). For example, one section of this installation examines the Western concept of race, suggesting that it is a largely political concept. In another area of the exhibition, visitors explore personal bias in order to reflect on the tendency of anyone (including those in power in our society) to allow personal points of view to cloud their perspective. Responses to these particular display areas were strong and often passionate. Roger, a student, described his experience as follows:

> If you go back to the genetic exhibit at the Science Center, that totally changed it for me . . . I didn't honestly think about it before, I was shocked

> when I saw that part of the gallery [that examines the concept of "race," revealing that it is largely a political and social concept with no firm basis in biology]. But yet, in everybody's heads, it's such a big deal. (Interview)

Comments from students suggest that exhibitions that attend to real-life connections and ground content in socially and personally relevant contexts potentially enhance visitor learning.

2. Evoking Emotion

Many have written about the role emotions, feelings, and attitudes play in visitor learning and behavior (Csikszentmihalyi & Hermanson, 1995; Dierking, in press; Falk & Dierking, 2000). Indeed, it has been argued that the chief impediment (and motivator) to learning is not cognition but affect. The role of affect is particularly important in science—a subject often described by students as "dull" and "boring." This is not surprising, as students wade through excessive content demands, usually void of context. Typically, science is presented as a corpus of knowledge to be mastered, memorized, and occasionally applied to the real world. However, critical issues-based installations offer something more than simple explication of scientific theories or principles. They strike at the very heart of controversy and debate, and inherently engage visitors by appealing to our intellect *and* our sensibilities.

In *Mine Games* many of the students became emotionally charged by the dilemma presented in the exhibition (although this is difficult to convey via text). These excerpts describe two different, but equally compelling, emotional responses to the possibility of a mine being built in their town:

> We know that this guy doesn't have enough money and he probably needs some more for his family, and so he needs the mine. (Chris, post–*Mine Games* interview)

> Think about it, we're dumping, 200 million tons of tailings into a lake. What if that's like some dad's secret fishing spot with his kid, and they go up there in the summer and all of a sudden they go up there one day and find this big mountain of green slime. That's kind of sad. (Mike, post–*Mine Games* interview)

A student visiting *A Question of Truth* was clearly moved by what she had seen in the "Bias in Science and Society" gallery:

> Certain people were considered inferior. When we went to the Science Center and they had that video on all those black people that were treated with . . . were given syphilis to research the causes and effect in terms of the disease. And they did that because some people, they just weren't considered as worthy as anyone else, it was horrible to see. (Ingrid, postinterview)

In another example from *A Question of Truth*, the emotional aspects of the exhibition were, ironically, considered to be too powerful or inappropriate for the audience. While designing the confinement box that worked as a metaphor for the coffin-sized sleeping quarters used in the forced transport of people from Africa to the Americas during the slave trade, curator McLaughlin (1998, p. 12) explains that "both within the institution and among our public, people were calling on us to stop the move to such a radical, 'real,' emotional experience." An earlier study conducted by Pedretti, Macdonald, Gitari, and McLaughlin (2001) found that the confinement box with its audio about the story of the slaves being transported was one of the most talked about (and visited) exhibit elements. Analyses of comment cards provided additional support for the strong emotional responses evoked by the exhibition (see Pedretti & Soren, 2003, for more detail). The passionate and often lengthy comments visitors wrote as they delved into issues of prejudice, racism, and truth were overwhelming:

> I am a member of the Metis Nation of Ontario and I have studied a little in Native worldview. Basically, my point is that ORAL TRADITION has been *shut out* by SCIENCE, STUDY, AND THE ALL POWERFUL WRITTEN WORD. I have participated in Pow wows and I have listened to many elders on matters relating to social "DISHARMONY"—My realization is that science can be a good thing if and only if it is balanced by the natural forces of Mother Earth. We need to slow down the pace of society and listen to the rhythms of Nature so that we can truly understand who we are and where we are going. Thank you so much for a wonderful exhibit and the opportunity to VOICE. (emphasis in the original)

Data strongly suggests that emotional elements of exhibitions (coupled with highly personal references) play a central and memorable role in visitor experiences. Evoking emotion potentially enhances learning by increasing engagement and motivation, thereby opening up possibilities for learning.

3. Stimulating Dialogue and Debate

Mediation through language is critical to learning. A focus on language promotes developing interest in reading critically and actively about science, competence in scrutinizing claims and arguments made in the press, and the ability to communicate complex socioscientific issues and/or positions effectively to others (Wellington & Osborne, 2001).

Both exhibitions uniquely position talk as important components of the visitor experience. Allan, one of the eighth-grade teachers, describes the virtues of dialogue, open-mindedness, and critical thinking:

> I think one of the most important things that I want the kids to learn, besides picking up a little bit of the actual science and observing things, is the importance of discussion. Get them involved in a discussion, a reasonable discussion and looking at the options, what are the options and then making a decision based on that and not just on hearsay or being so closed minded that "this is what I want." Look at the bigger picture, what else, and who else am I affecting by making this decision and how can I make that a little better? Can I make a compromise on that? (Interview)

One of the students, Roger, after visiting *A Question of Truth*, implied that hands-on exhibitions do not always stimulate discussion about science: "When you go to the other places around the Science Center, you've got experiments going on and you have the computers . . . actually, hands-on things, experiments. This was different. It made you think more and talk more" (postinterview).

If, as Koster (1999, p. 292) suggests, science centers are to serve as places for "airing of society's most vexing issues related to science and technology," then occasions to "air" visitor voices need to be provided in public spaces. Resolution of social issues requires opportunities for dialogue and debate (Wellington & Osborne, 2001). The need to converse, listen, and communicate has informed the direction of the *Mine Games* exhibit, particularly the *Hot Seat* experience. This innovative space encourages visitors to debate the merits of building a mine while considering social, economic, environmental, and historical impacts. Analyses of students' responses during the *Hot Seat* town hall meeting reveal rather sophisticated questions and concerns directed toward the facilitator (who mediates the discussion with school groups): "Mining destroys the land. I don't know about reclamation . . . they [companies] may not keep their promises about reclamation" (Neil), or "I think we need much more information than we have. There are lots of geologists, lots of environmentalists, and they all have different ways of showing what they're actually trying to convince you of . . . so I think we need to go to a few different geologists and environmentalists" (Katrina). In spite of the enormous success of *Hot Seat* with visiting school groups, it should be noted that casual visitors, although highly supportive of *Mine Games*, were less inclined generally to engage in public discussion through *Hot Seat.* Instead, they seemed to prefer engaging in more private debate with their companions. Perhaps participation in public discussion about potentially contentious topics, while visiting a science center, is too risky and beyond the casual visitors' expectations, or groups prefer to interact with one another rather than strangers when visiting museums.

These findings suggest that spaces for dialogue in nonschool settings can enhance the spirit of inquiry, allow for a free exchange of ideas, and encourage the formulation and articulation of carefully thought-out, defensible opinions. Subject matters found in these exhibitions inherently provoke discussion—they are intentionally chosen topics—reflective of multiple viewpoints. Furthermore, extensive scaffolding (i.e., through science center educational guides, mediators, and teacher preparation before, during, and after a science center visit) can enhance visitor experiences and meaning-making.

4. Promoting Reflection

Exhibitions like *Mine Games* and *A Question of Truth* cause visitors to reflect on the processes of science; the role of power, politics, and culture; and personal beliefs. This reflection can cause confusion and sometimes hostility, particularly when the sanctity of science is pressed (Macdonald, 1998). For example, 8% of the 50 people interviewed felt that *A Question of Truth* did not belong in a science center. One visitor in particular suggested that the exhibition was in fact "anti-science." However, 84% of the comment cards were overwhelmingly positive (see Pedretti et al., 2001)—applauding the science center's efforts to demystify and deconstruct the practice of science while providing a social cultural context.

It seems that reflection and dissonance caused by issues-based exhibitions can create powerful learning opportunities for visitors, as they struggle with multiple viewpoints and diverse value perspectives posed by the exhibitions. Consider the following quotes from students:

> The exhibit makes us think a lot about our beliefs and why we think certain ways. . . . I didn't think that the gene that effects the colour of your skins was so small and unimportant. Most people don't think of things like that. (Chris, student, postinterview, *A Question of Truth*)

> I think it opens people's eyes and we might have a bias about something, but the exhibit kind of gives you both sides to an issue and you really start to think, "Oh, well maybe I was wrong." Or maybe "I was right." But there's another way to go about it. (Joan, postinterview, *Mine Games*)

In the context of controversial subject matter, reflection inevitably leads to decision making at personal and/or societal levels. For example, students began asking where they would position themselves amidst controversy. What is my opinion on genetically modified foods? How do I participate as a citizen in public debate about such issues as genetically modified foods or reproductive technologies? What are the accompanying ethical considerations? While visiting *A Question of Truth,* students grappled with provocative and compelling issues: eugenics, slavery, intelligence testing, genetic engineering, and so on. In particular, they argued strongly for reflection, decision-making skills, *and* the need to exercise moral thinking:

> We view science as often being separate from morals, and it's kind of negative because it allows them to do all sorts of things like altering human life, and it may not necessarily be beneficial to our society. . . . Some scientists are saying, should we actually be doing this [genetic engineering, cloning]? And just because we can, doesn't mean we should. (Megan, student interview)

Megan's comment begins to explore complex questions related to what is and what ought to be, and speaks to the raising of social consciousness.

Examining one's attitudes and beliefs in the context of controversial subject matter is difficult. Deconstructing long-standing and often deeply entrenched personal views can sometimes lead to tension for visitors and staff alike, creating feelings of discomfort and unease. An interesting paradox emerges: the public expects issues to be taken up in settings like a science center, yet they also expect information to be provided and a "conclusion" to be reached (see, for example, Mintz, 1995). However, if we are striving to move away from simply presenting phenomenon or objects in glass cases, and enter into meaningful learning about controversial subject matter, then we must engage one another in critical thinking and reflection. Issues-based exhibitions provide those very opportunities.

THE DILEMMAS OF CHALLENGING CONVENTION AND COMMUNICATING CONTROVERSY

Tales to Tell and the Issue of Advocacy

Once a science center or museum has agreed to use issues-based installations, the thorny question of multiple viewpoints and advocacy rears its head. Should science centers or museums advocate particular positions? Should they remain unbiased, carefully presenting multiple viewpoints? Whose viewpoints? Can a "balanced" exhibition really exist? What is the relationship between the development of an exhibition and uncritical endorsements espoused by funders? Some issues are easier for museums to endorse—for example, energy efficiency, habitat protection, appropriate and sustainable technology, nutrition, and discouraging substance abuse. However, for some issues such as cloning, genetic reproduction, nuclear warfare, nuclear power, and genetically modified foods, there seem to be many more tales to tell—a multitude of perspectives—all riddled with complex moral and ethical repercussions.

In the case studies presented here, issues relating to advocacy, balance, and controversy played themselves out among science center staff, other stakeholders, and at times with the public. For example, *Mine Games* was heavily subsidized by a mining industry, causing concern amongst staff and the public. The exhibition was accused by some as being pro-mining rather than neutral in its presentation. What is clear from these examples is that the telling of narratives to the public requires balance and multiple viewpoints. However, at the same time, balance is not easy to achieve, and indeed some would argue that balance is impossible.

Mandates and Missions—What Do Visitors Want and Expect?

Any discussion about issues education in informal settings begs the question: How will visitors react? There is considerable controversy regarding visitor attitudes and opinions about issues-based programming and exhibitions (Cannon-Brookes, 1997; Mintz, 1995). Do visitors (the casual visitor or the school group) expect the "truth" at a science center or museum? Are they uncomfortable with representations of multiple viewpoints, open-ended exploration, value perspectives, or unresolved questions? At the *Mine Games* exhibition, the *Hot Seat* gallery was successful with students and teachers. However, casual visitors were reluctant to participate in public discussion about a potentially contentious topic and therefore the gallery was modified.

Issues-based exhibitions can inspire strong emotional responses, raising sensitive and profound questions. It is not necessarily "fun"—which is what the public often expects when visiting an informal setting. Yet, at the same time, the public agenda is shifting toward a reappraisal and critique of society and its relation to science, technology, ethics, the arts, and so forth, thus forcing museums to shift and reevaluate their role in public education.

Developing Issues-based Exhibitions

The two case studies referred to in this chapter are only a few of many innovative and thought-provoking exhibitions that have been developed in an effort to address contemporary issues. Many agree (Cannon-Brookes, 1997; Macdonald, 1998) that museums can no longer avoid hosting controversial exhibitions or installations. Arguably, these exhibitions are difficult to bring to fruition (for example, the team responsible for *A Question of Truth* worked and struggled for five years to gain and secure approval, conduct research for the content of the exhibit, and design and build it). Challenges to developing issues-based exhibitions include: (1) issues are not phenomena based, and therefore can be difficult to communicate with hands-on interactive exhibits; (2) few museums can provide staff presence at an exhibition, particularly one that addresses controversy; (3) subject matter is intrinsically difficult; and (4) conveying complex issues in an exhibition requires a significant amount of text (Mintz, 1995; Pedretti, 2002). How can we assist public forums in developing such exhibits? Although there are no foolproof methodologies a few suggestions and recommendations emerge. Cooks (1999) suggests that museums: (1) reach out to the community; (2) host a preview and invite potential adversaries; and (3) learn from the stories of people who have already hosted potentially difficult exhibitions.

CONCLUDING COMMENTS

What have we learned about learning through issues-based exhibitions? Findings suggest that critical exhibitions challenge visitors in different ways, they appeal to our intellect *and* our sensibilities. They evoke emotional, intellectual, and sometimes even spiritual responses in visitors. They provide experiences beyond usual phenomenon-based exhibitions and carry the potential to enhance learning by personalizing subject matter, evoking emotion, stimulating dialogue and debate, and promoting reflection. Thus issues-based exhibitions serve as excellent environments in which to further explore the nature of learning in these settings.

Working with issues-based exhibitions has also convinced me of the centrality of the "personal context." Although I highlight the personal in this chapter, the interplay amongst overlapping contexts (personal, sociocultural, and physical) cannot be overlooked. These innovative installations enhance learning through their increased attention to multiple contexts—not only visitor context, but the context in which the subject matter operates.

In summary, installations like *Mine Games* and *A Question of Truth* are deeply embedded in a rich social, cultural, and political milieu, and that is precisely what makes them so powerful and motivating, and for some, problematic. However, a number of recommendations can be put forward to assist museums as they begin to reinterpret their public roles, and the exhibitions that they create. Museums need to maintain an appropriate balance between past, present, and future in their galleries and public programs; find different ways of involving visitors in the subject matter; think carefully about the "tone of voice" in which they communicate with their visitors (i.e., thinking about credible alternatives to the scholarly, schoolmasterly, authoritarian voice); and be braver in their choice of exhibition topics and interpretive techniques (see Durant, 1996; Pedretti, 2002). Although a challenge to develop and mount, issues-based exhibitions promote healthy public debate about complex subject matter that touches all our lives—indeed, a worthwhile challenge to pursue.

ENDNOTES

The author is grateful to Vancouver Science World, the Ontario Science Centre personnel, and all of the teachers and students who so graciously participated in this research. Their unfailing support and commitment have made this work possible. The author would also like to thank the Social Sciences Humanities Research Council and the Centre for Studies in Science, Mathematics and Technology Education at OISE/UT for their generous support.

An earlier version of this chapter was published in *Science Education*, 88 (suppl.1): S34–S47, 2004, © Wiley Periodicals, Inc. Used with permission of John Wiley & Sons, Inc.

1. For more information on methods and specific research findings, please see Livingstone, Pedretti, and Soren, 2001; Pedretti, 1999, 2002; Pedretti and Forbes, 2000; Pedretti et al., 2001; Pedretti and Soren, 2003.

III

FOSTERING A LEARNING-CENTERED CULTURE IN OUR INSTITUTIONS

10

New Ways of Doing Business

Robert "Mac" West and David E. Chesebrough

This chapter identifies issues affecting museum sustainability and discusses strategies that can be used by museums to ensure their survival and success in the highly competitive world of the 21st century.

We use the term "business model," as defined by Falk and Sheppard (2006): "A business model is a description of the operations of a business including the purpose . . . , components . . . , functions . . . , core values . . . and revenues and expenses that the business generates." While many museums may be unaware that they have a business model, they in fact do and often it is in need of significant thought and repair or replacement.

We emphasize those organizations with which we are most familiar—science and children's museums—and look externally at models in other industries. Museums with other collections and perspectives face similar issues.

CURRENT CHALLENGES FACING MUSEUMS

Falk (2006) and Falk and Sheppard (2006) observe that the museum business model has changed as society moved from the Industrial Age to the Information Age and now to the Knowledge or Creative Age (see also Mallwitz, 2006). It is clear that the society within which museums now live is very different from that in which they emerged in the late 19th and early 20th centuries. For example, 75 years ago, the window on non-Western societies and natural history was, for all practical purposes, *National Geographic Magazine* and a few museums. Today it includes cultural and ecotourism, cable television, the Internet, large-format films, theme parks, festivals, CDs and DVDs, many new museums, themed restaurants and, for an increasing number of us, the family next door.

Bradburne (2004) assesses the relationship of (largely European) museums to public funding and their relevance to daily life. He suggests that the current approach to museums—educational venue and leisure attraction with benefit as an economic driver—is not sustainable. Further, simply building museums larger and architecturally more avant-garde, and filling them with expensive temporary exhibitions attractive to one-time visitors, does not make them more economically stable but, in fact, increases their economic vulnerability. An economic model based on ever-increasing attendance does not make sense. It is nonadaptive behavior—seeking a larger niche in a more competitive environment. Bradburne turns to the service model of the library, recommending that the museum "return to being a learning hub—not a 'destination attraction.'" This is consistent with the business strategy articulated by Collins (2001, 2005) and Collins and Porras (1994) as the "hedgehog concept." That posits that success in any business results from operating at the confluence of three circles—what the organization is best in the world at, what it is deeply passionate about, and what drives its resource/economic engine.

These and other commentaries, plus ongoing discussions at major conferences and symposia, suggest that we need to frame the conversation in ways actionable by the museum industry. This boils down to understanding the fundamental relationship of museums to their environment and their users/participants. Do museums understand and continuously respond to the ever-changing needs of the society in which they exist? Or are they a stable and predictable source of information and inspiration around which their society should coalesce? Regardless of where an institution falls on this spectrum, it must have a business model that allows it both predictability and flexibility, as well as effective mechanisms to continuously assess performance and make adjustments.

Several writers—in particular, but by no means limited to, Falk (2006), Falk and Sheppard (2006), and Weil (1999, 2002, 2005)—have commented on how the relationship between a museum and its community has changed over the past decades. There is a heightened expectation that a museum will be personally relevant, conceptually current, and open/responsive to (if not soliciting) visitor input.

Many institutions are struggling: they haven't matched their offerings with their desired or possible audiences; they are not perceived to be "important" or "essential" to their communities; there is a mismatch between revenues and expenses; and/or external circumstances have changed faster than the museums have responded. In other words, they are struggling to develop appropriate business models that fit them into niches actually available in the modern world.

BUSINESS MODEL FRAMEWORK

Chesebrough (2005), as part of ongoing Association of Science-Technology Centers (ASTC) discussions, developed a framework for analyzing the business models currently in use across the universe of museums. He is quick to point out that *there is no single model that applies universally.* Each institution adapts to its own particular set of conditions, sometimes through a gradual evolutionary process and sometimes through wrenching, rapid, unplanned changes. Absence of change may result in institutional failure.

Chesebrough's construct analyzes museum business models based on three questions:

- Why do museums exist?
- What is their role in meeting their community's needs?
- How do they provide their services in a financially sustainable fashion?

Bringing the "why," "what," and "how" together to create a sustainable approach to a museum's community is the current challenge and the basis for new business models.

Chesebrough analyzes several operating profiles that have run into sustainability issues, ranging from overbuilt newer centers that can't generate adequate support for their greater fixed operational costs to publicly subsidized older museums that find funding dramatically cut by struggling local governments.

We have seen conscious, aggressive diversification of museum revenue sources. In the 1980s and 1990s, there was a movement to add revenue-producing elements in search of a magic bullet that would resolve the unending problem of admissions-driven revenue failing to adequately support operations. The array of approaches has not achieved widespread sustainability. Simultaneously, some museum patrons/supporters question the "purity" of a revenue-driven focus. Museums are being challenged as selling out to the profit motive and violating their credibility with the public as nonprofit cultural and educational assets.

BUILDING RELATIONSHIPS WITH THE AUDIENCE

An important discussion, going back many years, well-articulated by Weil and central to Falk and Sheppard, involves the role of the audience/participants in determining the array of functions and services provided by museums. What guidance is being provided by current or potential users? How are museums deploying their resources—collections, exhibitions, facilities, staff, volunteers, finances, and so on—so as to best benefit their users? How much do museums respond to users, and how much do they assert their institutional authority to tell users what they "need to know"?

Case studies from the ASTC sessions indicate an increasing expectation on the part of the external community that their museums will be responsive to their expressed needs. Time after time, as West has conducted community environmental scans, he has found this to be the case. Two responses are heard again and again: "I don't go to the museum because 1) it hasn't changed since I was there last, and 2) it is not responsive to my personal needs."

The field responds to the first by instituting (often very expensive) changing/traveling exhibition programs and special events. Response to the second is much more difficult, as individual needs are legion, and there is only so much an institution can do to meet them all. An aspect of this is the need to stay relevant and current. The public's needs and expectations change, sometimes very rapidly. A successful museum (or any business for that matter) must stay current with society. An approach to this is to follow the Collins and Porras model (1994) and be introspective about their three-way intersection. In doing so, the museum attempts to serve a potentially narrower audience better, thus satisfying them and turning them from visitors into guests, then into participants and, ultimately, partners for life. Falk and Sheppard (2006) play out this model in the fictional Alabaster Museum of Natural History at the beginning of their book. But as the late Roy Shafer, an advocate of the Collins and Porras hedgehog concept of doing one thing very well, pointed out, too much emphasis on that, to the exclusion of a full range of activities, can result in a turning away of legitimate and needed revenue and losing the broad audience base needed to build deeper relationships.

SOURCES OF INSPIRATION

Where can museums go for inspiration for new business models based on close relationships with audiences/customers?

First, they can look at successful (generally for-profit) operations. Many have been brought forward and dissected in various forums. These include avant-garde libraries (Cerritos, California, and Seattle, Washington), independent bookstores (The Tattered Cover in Denver, Politics and Prose in Washington, D.C.), gourmet coffee shops (Caribou, Starbucks), quality groceries (Wegman's, Whole Foods), community art centers, upscale sporting goods stores (REI), and, very interestingly, megachurches. These organizations have established strong relationships with their customers/participants/members. Customers see value in their experiences—even when they pay premium prices, whether it be for a cup of Caribou coffee, lettuce at Wegman's, or Amazon-plus book prices at The Tattered Cover.

The reasons for success of each of the above vary and all can be looked to for some inspiration as museums look to create deeper, longer lasting relationships with their audiences. Wegman's knows that it cannot compete with Wal-Mart on food prices, so builds its business model around its motto of "helping you make great meals easy." Thus, it focuses on individuals who care about good food, are willing to pay for quality prepared foods and exotic foods from around the world (where the store has its highest differentiation and profit point), take free cooking classes from the in-house chef at every store, get recipes from the company's website, and have a pleasant in-store experience served by staff that have won the company high rankings in surveys of best companies to work for.[1]

Starbucks has made itself accessible through a vast network of outlets, has created an environment that helps replace the "third place" that some feel has been lost (with the societal demise of the old neighborhood lodge, bar, or bowling alley; Falk & Sheppard, 2006), and has increased revenue per customer by providing the opportunity for the coffee-sipping customer to listen to 150,000 songs and create, burn, pay for, and walk out with a personalized CD in addition to the price of the coffee. Further, Starbucks is an "aspirational environment," one in which people are both ready to be seen and where they know they will associate with similar people.[2]

Another instructive example is megachurches (the 1,000+ U.S. congregations with weekly attendances of 2,000 or more; *Economist*, 2005). Willow Creek Community Church of South Barrington, Illinois, bases its program on a survey of the target audience of nonchurchgoing suburban Chicagoans. As a result, Willow Creek has no overt religious images (crosses or stained glass); services that include videos, drama, and contemporary music without over-the-top evangelism; and impeccable grounds with convenient parking.

Even though they are huge, megachurches maintain personalized relationships through small groups with similar profiles. Willow Creek hosts affinity groups ranging from motorcycle enthusiasts to those watching their weight. It provides child care on weekends so parents can attend services. The churches segment their market, providing different services at different times with different emphases for different audiences.

The organizations or businesses mentioned above succeed at creating a personalized, customer-satisfying approach that has translated to a successful "brand" by which they differentiate themselves and encourage repeat use by a loyal customer base.

Second, museums can look at their colleagues who are changing existing organizations and establishing new ones with new sets of practices, policies, and community relationships. Many museums are developing creative programs, leveraging their resources in new ways, and extending themselves to new and existing audiences in exciting and provocative ways. We believe that there are potentially successful new business models that are being actively tested, are works in progress, and are on the drawing boards. Comments on three of these follow.

BUFFALO MUSEUM OF SCIENCE

The Buffalo Museum of Science (BMS) is an institution founded in 1861 and heavily government subsidized for much of its history, primarily functioning as a traditional, but sometimes innovative, natural history museum. The declining upstate New York economy came to a crisis point after 9/11. Quick and significant cutbacks in government support accelerated a change process that was already underway at BMS.

A team of national experts joined then–BMS President Chesebrough and his management team in looking at the core mission of the museum, the environment in which it was operating, and the best purpose it could uniquely and passionately pursue.

Purpose moved to focusing on being an essential educational anchor and partner in its inner-city location and providing a venue and avenue for lifelong learning. Research and collections maintained their valued place as unique assets, but now largely as powerful tools to help engage and inspire learners of all ages.

At its core, the new approach is about moving occasional visitors to regular users and then to lifelong learning partners. In the process, those who choose to move to greater depths of relationship will produce more revenue per person over time.

Approaches to service changed dramatically with the new purpose. The new strategy focuses on partnerships and more in-depth, layered offerings at higher price points (and greater net revenue). This approach of greater synergy and alignment of programs and resources is also tied in with relationship-based fund-raising leveraging the museum's impact. There is more focus on multiyear personal donor gifts to general operating support and targeted endowment building to supplement earned revenue growth. This will replace the declining government and corporate support and ensure a core level of access programs for inner-city families.

To support this physically, the classic 1929 building and its traditional, unchanging diorama-based gallery experiences (supplemented by traveling exhibits) is envisioned to become a vibrant, community-based "hub" for a system of contact points for the public, and more specifically, the learners who develop an interest in the world around them.

Key changes in building and program design have included:

- *Connections*—an experimental inquiry-based space that replaced a 4,500-square-foot static exhibit hall—became the showcase test element for an open-ended, facilitated lab space. It is object-rich with both items from the collections and touchable objects clustered at activity stations complete with exploratory devices ranging from magnifying glasses to computers for digitizing microscopic images. Staff has become adept at quickly changing topics to match current news topics or allow for a varied experience from week to week.
- *Authentic Learning Communities*—BMS staff developed educational packages (sold to schools as contracts) that build on teacher workshops, multiple student engagements, and interactions with scientists and staff educators, accessed partly through outreach.
- *Youth and Neighborhood Partnerships*—a concerted effort, underscored by hiring a community relations manager—led to many new relationships and access programs with neighborhood community groups and schools, resulting in programs weaving a web of ongoing contact with inner-city youth and their families. The museum has been told that it has become the "de facto" center for activities of the African American community.
- *Science Spots*—The museum is experimenting with going to where the audiences are. A first storefront operation has been opened up in a vibrant city neighborhood. The Science Spot offers neighborhood-specific programming, meeting spaces, and family and children's activities. While not a substitute for the main museum, the small program and retail operation offers convenience and a stimulus to go to the main museum for deeper and richer experiences.

Operational and organizational changes were made to use declining resources most efficiently and to strengthen the institution within a web of partnerships. Key changes have included:

- A variable schedule that matches with core audience's availability and has the museum closed to the public during slow periods.
- Shift to a higher proportion of science educators and careful hiring for passion and skill in using the power of real objects and inquiry-based learning approaches.

- Enhancements to scientific services such as research and conservation care, using the skills and interests at local colleges and among local interest groups.

Whether the capital funds to move the full model forward will be available in the near future is still in question in a struggling Western New York economy. But the BMS now has a staff passionate about service; a new relationship-based model providing more in-depth, layered learning opportunities is in place; it is making the best of the investments made in the museum and the Science Spot; and new partnerships and relationships are growing out of the new approaches.

In the process, donor-giving growth for operations and endowment, and the ability to charge higher fees for a differentiated experience, are starting to change the economic model of the museum for the better despite the declining government support.

CHILDREN'S MUSEUM OF PITTSBURGH

The Pittsburgh Children's Museum weathered the exodus from its North Side location, including the 1991 move of its next-door neighbor, the Buhl Planetarium and Science Center, to its new location as the Carnegie Science Center on the Ohio River. Starting in 1999 with leadership of Executive Director Jane Werner, a new vision and focus emerged that was unusually large for a small children's museum in a 20,000-square-foot landmark building. It leveraged not only a relationship with a children's icon, Fred Rogers, but the outsized thinking and ambition to serve children as well as have a broad community impact.

Werner and her team saw the potential to create community good in several ways that gave the museum purpose not only as a different kind of children's museum but also as a force for urban redevelopment. The approach they took, and the product that stands today after the November 2004 opening of the renamed Children's Museum of Pittsburgh (CMP), is a model of core purpose driving and succeeding through collaborative planning, development, and operation.

CMP starts with "Play with Real Stuff," an educational philosophy grounded in solid research. Werner and her team are driven by respecting the formative time of childhood so that children are given the opportunity to play and explore their world. To implement that approach they added the dimension of a partnership-based "town square" for child-centered organizations and blended the two "big ideas" into a design and business model.

From an experience standpoint, CMP offers:

- Exhibits designed for process, multilayered and appealing to a broad age range, families as well as children.
- In-house exhibit development, supported by a partnership with the University of Pittsburgh Center for Learning in Out of School Environments (UPCLOSE) that allowed for extensive prototyping of exhibits.
- "Real stuff" programs that include woodworking, simple machines, artistic endeavors of a wide range, all supplemented by regular demonstrations and interactive learning activities by artists and others.
- In-house partners with shared attention to children and families include not only the UPCLOSE research office but also Reading Is Fundamental Pittsburgh, Pittsburgh Public Schools Pre-K/Head Start programs, Child Watch of Pittsburgh (an at-risk children's agency focusing on education, advocacy, and programs), and the Saturday Light Brigade (a family radio program broadcasting from the state-of-the-art studio in the Children's Museum theater).

The notable architectural and neighborhood development approach, which supplements and contextualizes the visitor and child experiences, includes:

- Creating a children's campus that linked two landmark historic buildings with a signature entrance and exhibit space, with an exemplary commitment to both architectural integrity and environmental stewardship that earned the facility a Silver LEED rating and an AIA National Honor Award (a first for a children's museum).
- Community building through formal partnerships with child-based organizations that rent space and operate within a cost-effective "incubator" space concept. This has resulted in the museum and its partners working side-by-side, sharing ideas, resources, and creative energy to improve the lives of children and families.

- Partnering with other nearby institutions to continue to expand the "children's campus" approach to neighborhood economic development that now has a collective reopening of a neighboring theater.

The business approach, driven by their core purpose and philosophy, has resulted in not only a model institution but a healthy one. The power of their "big ideas" and the demonstrable and nationally noteworthy approaches have resulted in:

- Attendance growth of 162% over the previous facility, exceeding the highest expectations.
- A capital campaign of $28 million that was exceeded by almost $1 million.
- Endowment growth from $1 million to $6.5 million as a result of the strong appeal of the project and its capital campaign.
- Increased earned revenues across the range of admission, retail, programs, food, and parking services.
- A notable ability to raise grant funds through the built-in partnerships and research-backed efforts.

ONTARIO SCIENCE CENTRE

The Ontario Science Centre (OSC) has been a leader in the development of hands-on visitor engagement since its opening in 1969. Its evolution paralleled that of the Exploratorium, but with a markedly different style, esthetically and educationally. It has become the most-visited museum in Canada.

By the turn of the century, many of the experiences that initially were unique to science centers—and the staple of the OSC experience—were available in numerous other venues. Interactive experiences now can be had at venues ranging from restaurants and retail to the Internet and cable TV. While the OSC was not yet seeing decreases in either audience or revenue, it decided to preemptively respond to what it saw occurring elsewhere in the science center field. Thus, starting in 1999, guided by then-new Director General and CEO Lesley Lewis, OSC began planning proactive changes.

The OSC carefully researched both its visitors and its supporters. It found a high comfort level with the center as a whole, but learned that it was perceived as largely a children's experience. Teens and young adults (13- to 24-year-olds)—who were making vital career and life decisions, those that determine who and what they will become—were noticeable in their absence as OSC visitors.

As the center looked beyond itself, the world's focus on innovation and creativity as a prerequisite for increasing global competitiveness became apparent. It discussed strategy options with leaders and practitioners in education, research and development, and the corporate sector. The OSC then developed a new model for visitor engagement. This seeks to move from having an impact solely on the casual visitor to building relationships with people of all ages and extending this relationship beyond the visit to the OSC site into other aspects of individuals' lives. It continues the focus on science and technology, but broadens the array of both disciplines and experiences provided by the OSC. Further, it emphasizes the concept of problem-solving as the way to provide its guests with the skills and attitudes for addressing current science.

Much of OSC's thinking is inspired by Albert Einstein's statement, "The problems that exist in the world today cannot be solved by the level of thinking that created them."

The OSC now is changing the relationships with its visitors, transforming them from occasional users to participants, from consumers of science "facts" to participants in the process of science. It is challenging itself to help create the corpus of innovative individuals who will lead in the solution of the challenges of the future.

There are new relationships with supporters—OSC is not just asking for money, but now is asking to become partners in developing the experiences. For example, DuPont Canada is a "Knowledge Partner" who participates in an exchange of knowledge and staff expertise with OSC. The MIT Media Lab and the Ontario College of Art and Design collaborate on the Media Studios. Ryerson University's Early Childhood Education Program provides intern students to be hosts in KidSpark and earn academic credit.

The center now has a staff position of Partnership Coordinator. He uses a protocol for assessing the viability of a proposed partnership, which includes a pilot effort, completing a Compatibility Matrix Template, careful management of the relationship, continual evaluation of partnerships, ensuring that senior management on both sides communicate and support the relationship, and having an exit strategy.

The mission of the initiative is to spark creativity, inspiration, innovation, and change by joining participants and partners with the Science Centre to create unexpected experiences, relationships, networks, environments, and enterprises so that people generate new ways of seeing and thinking about themselves and the world.

Goals have been established that include increasing public awareness of the importance of innovation and its processes and products, providing access to current science and technology research and discoveries, introducing young people to career opportunities in science and technology, revitalizing the OSC and continuing its leadership on the world stage, and creating new, deeper, and ongoing relationships between visitors and the Ontario Science Centre.

The expanded mission of the center requires a physical platform for change—over 30% of the center's public space (42,000 square feet) is being transformed into new environments. It also is creating a 55,000-square-foot outdoor science park at its entrance. The first new area—*KidSpark*—opened in 2003 and doubled in size in 2005. Part One of the Weston Family Innovation Centre (the *HotZone*) opened in March 2005 and the balance of the Innovation Centre (20,000 square feet) opened in July 2006; the Exploration Plaza opened in August 2006 and three major art installations were completed in November 2006.

The current initiative is a work in progress. It initially was estimated to be a $30 million effort in 2002; by 2006, $47.4 million had been raised and the scope increased dramatically. Additional planning is projecting further changes in the array of experiences with longer-term objectives to significantly increase both on-site and off-site use of the OSC.

CONCLUSION

It is clear that the world in which museums live and operate is changing, and many established operating models are no longer adequate or appropriate. Thus, it is necessary for museums to be focused on the intersection of mission, economics, and a changing society.

Numerous nonmuseum organizations, both for-profit and nonprofit, provide (at least partial) models for the development of new, intense, and often revenue-positive relationships with customers/participants. A growing number of museums are also experimenting and implementing new service and relationship models that deliver more personalized services, create new community partnerships, and change the business model from heavy reliance on infrequent and new visitors to deeper relationships with a more highly engaged public.

We are confident that more museums will monitor and learn from these experiments and models and will follow up with still more, hopefully successful, and effective 21st-century museum business models.

ENDNOTES

Most of Chesebrough's work on this chapter was conducted while he was president and CEO of the Buffalo Museum of Science, Buffalo, New York.

We appreciate the invitation from John Falk, Institute for Learning Innovation, to prepare this chapter, and acknowledge assistance from Goery Delacote, formerly of the Exploratorium and now of At Bristol; John Jacobsen of White Oak Associates; Alphonse DeSena, formerly at Exploration Place and now with the National Science Foundation; Chuck Howarth, Maeryta Medrano, and Don Pohlman of Gyroscope; Lesley Lewis of the Ontario Science Centre; Carroll Simon of the Buffalo Museum of Science; Thomas Krakauer, formerly of the North Carolina Museum of Life and Science; Jane Werner of the Children's Museum of Pittsburgh; the late Roy Shafer; Dan Martin and Tom Moriarity of Economics Research Associates, and many others who have provided us with information, food for thought, and challenges to our assumptions.

1. These comments are based on www.wegmans.com, supplemented by Chesebrough's conversations with store officials.
2. Thomas Moriarity of Economics Research Associates, Washington, D.C., personal communication to West.

11

Optimizing Learning Opportunities in Museums: The Role of Organizational Culture

Janette Griffin, Lynn Baum, Jane Blankman-Hetrick, Des Griffin, Julie I. Johnson, Christine A. Reich, and Shawn Rowe

The organizational structure of a museum has a direct impact on the types of educational experiences it creates, and therefore ultimately on visitor learning itself. While numerous books and articles have been written that describe the impact that the design of educational experiences has on visitor learning, few, if any, have explored how a museum's organizational structure influences the educational experiences it creates. This chapter will explore different theories, models, and research findings (from both within and outside the museum field) that describe what it means to be an effective museum, one that develops programs which provide enriching experiences for the visitor.

WHAT IS AN EFFECTIVE ORGANIZATIONAL CULTURE?

Museums, like other effective organizations, should make a difference to people's lives. However, how one assesses the "quality" of that experience is a critical and extremely contentious issue. Effectiveness is a value-based and time-specific construct and a political rather than scientific concept. Though outcomes should be related to the policy framework and objectives of the organization, what constitutes effectiveness often depends on whom one asks: different constituencies have different views depending on their relationship with the organization.

All organizations are groups of people seeking common outcomes through their involvement, outcomes that are both personal and that benefit the organization. There may be differences in complexity of the tasks and goals may be expressed differently, but most employees share views of what constitutes a decent place to work (Heintz, Wicks-Lim, & Pollin, 2005). Many of those views coincide with the characteristics of effective organizations. Leaders are firm but fair, staff know when they have done well and are provided opportunities for personal development, and values expressed by managers are those that are used to deliver rewards including promotion. The workplace is where safe and honest communication that respects others is valued.

LESSONS FROM THE FOR-PROFIT WORLD

Although many investigators of nonprofits have objected to being compared to the for-profit field, there are lessons the nonprofit field can learn by examining research from the for-profit sphere. Studies of successful for-profit enterprises show that contributions to effective outcomes are a strong organizational culture, leadership that focuses on development of staff potential and investment in training and development (i.e., "capacity" building), and recognition of contribution (i.e., "purposiveness"). Of critical importance to success is how people work together and how decisions get made (Hout & Carter, 1995; Lublin, 1993).

Four features typify effective for-profit enterprises: (1) They focus on vision and core values and build a culture to support these (Collins & Porras, 1994). They are able to envision the future through knowledge of industry trends and relate these to the organization's work (Hamel & Prahalad, 1994). (2) They focus on people: recruiting new staff, career development, promotion from within, and work force satisfaction. High-performance work practices also include contingent compensation, employee participation, higher wages, and reduction of status differences (Pfeffer, 1996). (3) They focus on teamwork (Guzzo & Dickerson, 1996). And (4) they build a learning orientation (Kofman & Senge, 1995).

Supporting a major change into a learning organization involves disrupting the modus vivendi. People's natural inclination is to feel threatened and to avoid anything uncomfortable or unfamiliar. To minimize these obstacles, Heifetz and Linksy (2002) suggest these techniques:

- Operate in and above the fray: One must maintain perspective in the midst of action; one must continually step "outside" to observe the situation as a whole, and then move back into the situation.
- Court the uncommitted: Recruit partners who can protect and point out potential flaws in the initiative or approach.
- Cook the conflict: Let tension exist. Conflict is a necessary part of change, but the key is to manage people's passionate differences to harness their energy.
- Place the work where it belongs: Leaders need to resist the need to provide all of the answers. Let the work be owned by all.

In museums in which change has been managed effectively (Abraham, Griffin, & Crawford, 1999)—where consequent performance of the museum was judged by staff to have improved as a result—the museum's management was able to translate external needs to internal vision and then craft appropriate actions by staff which integrated tasks, structures, processes, and systems at the technical, political, and cultural levels. Time and resources were allocated to the change process and the advantages of the changes were carefully communicated to key internal groups of staff.

Of great relevance to museums are the studies of creative organizations and ones engaged in research and development. In the successful ones, there is a continual challenge of ideas, and exposure to other views and current knowledge through discussion within the enterprise and between similar organizations (Coglisera & Brighamb, 2004; Mumford & Licuanan, 2004). Moreover, leaders promote a clear vision, provide intellectual stimulation, and encourage social contact—no hierarchies. Critical evaluation of work is also encouraged.

Many studies support these findings and they are all relevant to museums and other nonprofits. None of them emphasize efficiency or conformity as critical to success and all of them point up the importance of attention to people and their development, *no matter what kind of organization we are talking about.*

LESSONS FROM THE MUSEUM WORLD

Many expositions on how museums might be successful are anecdotal or are accounts of a single museum. Some are histories, such as that of the

Metropolitan Museum of Art in New York (Tomkins, 1970). Some are shorter accounts of change in a single museum, such as that by Janes (1997) of Glenbow in Calgary, Alberta. There are other interesting accounts of change at the Children's Museum in Boston, the Exploratorium in San Francisco, and elsewhere given by Gurian and others (Gurian, 1995). McLean (1999) gives a more recent account of how Exploratorium staff approached the challenge of improving public experiences without losing the essence of that place, and Emery (2002) recounts the challenges of reforming the Canadian Museum of Nature in Ottawa. Few of these reports relate performance to *relative* success: How do we know that the practices described are likely to lead another museum to success? The important contribution made by Weil (2002) to this area by his emphasis on "purposiveness" and "capability" as critical features of a successful museum, more so than efficiency or even effectiveness, deserves more attention in the museum literature.

The criteria for successful institutions proposed above are echoed in the findings from one of the most extensive studies of effective museums conducted to date. This study (Griffin & Abraham, 2001) examined 33 museums of all kinds, including science centers and aquaria, in Australia, Canada, the United Kingdom, the United States, and New Zealand (one museum only). Two sets of data were collected. Using an established procedure (Murray & Tassie, 1994), a set of experts (people considered to be experienced practitioners in the field) were asked to assess the effectiveness of a range of museums and provide an indication of the extent to which research, collections, public programs, and marketing contributed to those assessments. The experts rated the museums with which they were familiar. The expert assessments were surrogates for effectiveness, direct comparisons of museums in many different domains with respect to attributes such as financial performance and attendances being highly suspect, and visitor ratings of the museums' public programs beyond the resources of the investigators. Secondly, the staff of those museums responded to a questionnaire on management processes and behaviors in their museums, for instance, the extent to which goals and objectives of various sections are cohesive and well integrated with those of the organization as a whole, or the extent to which a learning orientation is encouraged. Respondents were *not* asked to make judgments about whether those processes and behaviors were good or bad. The expert assessments and the questionnaire responses were compared.

Amongst the almost 30 items that were found to characterize the museums rated by experts as most effective were a concern for quality, shared goals, good communication, attention to training, and strategic allocation of resources (Griffin, Abraham, & Crawford, 1999). In effective museums, senior managers work together as a team, staff actively support the goals of the museum and demonstrate a high degree of commitment and involvement, departmental goals are well integrated, and everyone is encouraged to respect the skills and contribution of others.

Public programming in effective museums emphasizes strategic approaches to achieving positive outcomes for visitors, including provision of a variety of learning strategies, ensuring that exhibits are in working order, and attending to problems "on the floor." The first of these recognizes constructivist approaches to learning, the visiting experience being much more than just inspection of the exhibits but rather an opportunity for a further elaboration of one's understandings. These characters merge into two key factors: cohesive leadership and visitor-focused public programming. It is not just leadership that is important, it is leadership that encourages development of shared values, supports a commitment to agreed standards of quality, and constantly encourages effective communication—leadership that provides opportunities for training, and rewards superior performance in terms of agreed and understood standards.

EFFECTIVE ORGANIZATIONS AND THEIR IMPACT ON LEARNING: CASE STUDIES FROM THE FIELD

In order to create an environment that ultimately will lead a museum to become more effective, some self-assessment needs to take place. The features described above as best for effective for-profits (vision/core values; people-focused; collaborative; building the learning environment) and those typifying effective museums (cohesive leadership; visitor-focused) are not mutually exclusive. Our ability to translate these features into all levels of our organizations is key to building a museum as a learning organization.

It is also important that whatever system is created be dynamic. Phillips and Case (2005), in an article entitled "Museums Grow Old and Calcify," caution the field against favoring and staying with the "getting by" agenda over working on the "getting better" agenda. The former is filled with the daily things that keep us busy and, to a large extent, challenged by the mundane. In contrast, the latter necessitates resolving problems at their root; designing an organization that works for, not against, staff; harnessing and nurturing energy; and surfacing and challenging assumptions. The authors further posit that the traditional hierarchical structure of museums works against them becoming learning organizations. Hemp and Stewart (2004) speak to the need for an organizational system in which staff not only deeply understand the mission or vision but are also empowered at all levels to make decisions and generate discoveries. In both of these writings, leadership is not top-down but is (and should be) distributed throughout all levels of the organization.

The critical questions are: How can these techniques be put into action? And if they are put into action, what is the impact on the visitor experience? The following section illustrates some specific examples in a range of museums and circumstances.

Conner Prairie

Conner Prairie is an open air "living history" museum located in Fishers, Indiana. Its 2004 mission statement, to "inspire curiosity and foster learning about Indiana's past by providing engaging, individualized and unique experiences," is enthusiastically embraced by staff and is evident in programming and visitor experiences. This was not always the case. By the late 1990s Conner Prairie was beginning to "grow old and calcify" (Phillips & Case, 2005); attendance had stagnated. Marketing studies indicated that while most respondents had fond nostalgic memories from a childhood visit to Conner Prairie, they felt there was little to motivate a return visit (Strategic Marketing Research, 2000). Interpretive staff used many of the same live interpretation techniques that had been used since Conner Prairie's inception in 1964.

Disruption in the modus vivendi began during summer of 2000, with the first in a multiyear series of research studies on family and interpreter interaction. Under the auspices of the Museum Learning Collaborative (MLC), Rosenthal and Blankman-Hetrick (2002) found that interpreter presentation style had an impact on a family's learning and communication about what they were experiencing. Interpreters using the traditional monologue presentation style and concentrating on delivery of historical information offered little engagement and involvement to families. Interpreters who were dialogic in nature captured the attention of children and parents and stimulated family conversation about the experience. In a separate study looking at visitor conversations, MLC found that the interpreters monopolized most conversations, leaving little room for interfamily conversation or interaction (Leinhardt & Knutson, 2004). This research, coupled with active pursuit of visitor feedback, prompted Conner Prairie's management to initiate changes in its practices. A Family Learning Task Force was initiated and given the challenge to develop a more "family friendly" interpretive plan.

As is common in many organizations, change was met with resistance. In this case, change demanded a different museum culture. This meant redefining what Conner Prairie was about and creating a common value system that was apparent in all aspects of organizational communication. Ken Bubp, special projects manager for Conner Prairie, stated,

> Despite their best efforts, managers found themselves unable to help and encourage individual interpreters to alter their approaches to interpretation; many staff continued to operate under long-held assumptions and struggled with how to make the sort of changes now being asked of them. Some wanted to change but did not know how; others simply did not see the need for change. (Seig & Bubp, 2006)

In response to this dilemma, *Opening Doors*, an initiative created by Director of Museum Programs Dan Freas, was introduced in the spring of 2003.

Through this new initiative, a small group of frontline interpretive staff were empowered in a team setting to examine the visitor learning and create the most effective learning environment. Conner Prairie's Administrative and Historical Programs management gave the team the freedom to think broadly and creatively within a certain set of boundaries. Here are a few (Seig & Bubp, 2006):

- Visitor learning experiences will serve as the touchstone for any efforts. Engaging visitors is the primary lens through which all interpretation should be seen.
- Experiences developed and provided by staff should reflect current understandings of visitor research at Conner Prairie, and of learning theory.

- The physical environment and historical content should be at the highest level of historic authenticity.
- The safety of guests is an essential consideration for any learning experience.

With the backing of the president, Ellen Rosenthal, a Visitor Focus Seminar was created to educate the team in visitor demographics and visitor motivation for attendance. Created by mid-level managers, the seminar gave special emphasis to ongoing research on how interaction between interpreters and visitors affected learning. Staff were also informed that a follow-up to the 2000 learning research study would take place in 2004.[1]

Institutionwide, there was a shift in how Conner Prairie viewed the visitor. Visitor experience reports and visitor feedback started being distributed to all staff members on a monthly basis. "Experience management" was a topic of discussion across the organization in staff meetings large and small. Rosenthal sought to close the gap between theory and practice by making the visitor experience, learning theory, and the quality of service part of the everyday conversation at Conner Prairie. Ongoing visitor research and employee feedback show these changes taking place. Conner Prairie was awarded an IMLS[2] National Leadership Grant to share this new approach in creating engaging museum experiences with the rest of the museum community. A frontline staff training DVD , Opening Doors to Great Guest Experiences, now available.

Museum of Science, Boston

For almost 20 years, the Museum of Science in Boston (MOS) has been actively working to design exhibitions that are physically, cognitively, and socially inclusive of persons with disabilities. While the museum is now recognized as a leader of inclusive design in the broader museum community, it was not that long ago that the museum was inaccessible to a large portion of its visitors. What led to this change was a recognition by senior management that ideas generated at all levels could have a profoundly positive impact on visitor learning.

In the early 1980s, the museum displayed a small exhibit that included artifacts from the life of Helen Keller. There were a number of items behind glass in the display case, including her famous typewriter. Groups of students from local schools for the blind began arriving at the museum and they wanted to touch the artifacts, especially the typewriter. MOS was not prepared for this audience and no thought had been given to making any part of the exhibit tactile.

Today, the Museum of Science is a very different place. All kinds of learners, including people with or without physical disabilities, are provided equal access to the museum's exhibitions. Exhibits are tactile and include carefully placed audio labels that provide increased access for visitors who are blind or dyslexic. Written labels utilize visually clear fonts with high contrast backgrounds. Wheelchairs and motorized scooters can easily access almost all the public areas of the building. The driving force behind much of this change began with the efforts of one key staff member—Betty Davidson, a wheelchair user who began her work at the museum as a volunteer after already having a full career as a biochemical researcher. This work is continued today by Christine Reich, manager of research and evaluation at the museum.

Much of what drove the initial changes stemmed from the vision of a committed staff, combined with some "accidental opportunities" that helped bring the rest of the museum staff along. This matches what we know about the adoption of inclusive design by for-profit companies; most are not driven to make products that reflect inclusive design because they believe they will be profitable but because of the personal commitment of key personnel (Trace Center, 2004). As an advocate for persons with disabilities, Davidson had the challenging job of creating interactive and hands-on components for the museum's large-animal dioramas, amid much skepticism that anything could or should be done. In the course of evaluating the resulting exhibition, an unexpected outcome was revealed. Not only did the modifications provide access to the experience for persons with disabilities, but the exhibit was also improved for visitors without disabilities. The percentage of visitors who had seen the exhibition and could name one animal adaptation went from 20% before modifications to 100% after. The number of visitors to the exhibition, the duration of their stay, and the amount of conversation they engaged in also increased dramatically (Davidson, Heald, & Hein, 1991).

The "accidental opportunity" of finding that accessible environments can improve the experience for persons without disabilities, as well as those with disabilities, was a pivotal moment in the push to change the organizational culture. Suddenly, the concept of inclusion made sense to all levels of the museum. Larry Bell, vice president of exhibits, instituted a mandatory policy that all new exhibits were to be reviewed for accessibility by Davidson prior to final production (Bell, 2000). The National Science Foundation was so impressed with the evaluation results that it funded publication of the study (Davidson, 1991). Consequently, those staunch internal supporters of inclusive design now had some credibility and support.

A new approach and way of thinking has become embedded in the organization, illustrated by the fact that neither Reich nor Davidson has to be in the room for inclusive design to be part of the conversation. Exhibit teams now voluntarily ask persons with disabilities to review and test new exhibit designs without prompting or directing from above. In recognition that it is not enough to "get by" and there is always a need to "get better," the museum has recently launched a new research initiative dedicated to the study of inclusive design, particularly focusing on the development of new methods for creating equitable learning experiences for visitors with and without disabilities.

It took years of diligent advocacy for these changes to take place. Today, a broader spectrum of people have ownership of inclusive design. In addition, visitor learning continues to be positively affected by the focus on inclusive design. A recent study conducted by Reich (2005) demonstrated that design features added to exhibitions for visitors with specific disabilities were appreciated by and aided learning for visitors without such disabilities.

CONCLUDING THOUGHTS

Key elements in the cycle of change that were illustrated through the above cases include the need for (1) a committed individual or individuals to be the standard bearer, (2) multiple strategies for engaging the "nonconverted," (3) time and patience, and sometimes (4) an external validator of the internal change.

In each case, major changes in the organizational culture of the museum had a direct impact on visitor learning. This is not entirely surprising—the learning environments that are constructed shape the kinds of learning that can be done there and the kinds of learners who are welcomed and supported. In fact, the kinds of changes that characterize a successful learning organization at the management and daily practice levels reflect some of the most important tenets of both constructivist and sociocultural approaches to learning—both of which are embraced by a wide spectrum of educators and learning specialists in the museum field.

Sociocultural perspectives on learning highlight the need for promoting experiences that activate visitors' funds of knowledge as a way of opening the floor to a wide variety of participants and leveling the traditional hierarchies of education. When staff are committed to including multiple sources of expertise in their practices and exhibits, they must begin to recognize that expertise comes from outside the walls of the museum as well and can be activated as a powerful motivator for, and tool of, visitor learning. In addition, recognizing what staff or visitors know is a way of activating prior knowledge. From a constructivist perspective, this is a crucial first step in learning since new knowledge and practices are built on old knowledge and practices.

This inclusion of multiple perspectives also recognizes expertise as something that emerges from practice. What is valued gets shifted from "experts" to "communities of learners" or "communities of practice," where knowledge, skill, and authority for decisions and meaning making are distributed among both novices and experts in the community (Rogoff & Lave, 1984; Wenger, 1998). When visitors are invited into the museum as members of its community of practice, rather than recipients of its expertise, the museum shows a commitment to expanding the number and types of people who are involved in determining what counts as a good, high-quality experience. The results can run the gamut from making room for one person to make a more meaningful connection to his or her prior experience and knowledge, to creating welcoming space for groups who never saw the museum as a welcoming or valuable space before.

Constructivists stress promoting activities that ask visitors to challenge assumptions and received knowledge, as part of the accommodation of prior knowledge and assimilation of new knowledge that contribute to cognitive change. This "metacognition" is recognized as an important element of learning for all ages. When a museum values self-criticism and open self-evaluation, it can model the process for visitors and create opportunities for visitors to evaluate their own learning and, in turn, learn to critically examine the museum as a source of knowledge or practice.

While some museum professionals may see this as shirking responsibility, Roberts (1997) describes it instead as a way of empowering visitors as critical meaning makers.

Sociocultural and constructivist approaches to learning also stress the importance of recognizing that learners come with and may value multiple and different learning styles and strategies than those that are promoted in the traditional museum setting. Just as in the story of Helen Keller's typewriter above, an emphasis on the preciousness of historical artifacts and the favoring of curatorial voice over affective experience supports some learning styles and strategies but actually locks out others. When the museum is committed to learning in all its cognitive, social, and affective aspects, the very definition of what it means to interact in a meaningful way with an object may change. As the examples above illustrate, recognizing the power of this change for one group of learners can improve the experience for all learners. When organizational culture is in line with what we know makes for valuable and inclusive visitor learning experiences, then the possibility arises for cross-fertilization of ideas and practices between the museum as an organization and the communities of visitors it serves.

It is proposed that effective exhibitions are developed by museums in which these features of effective working relationships and commitment to an agreed method of decision making are pursued. The hierarchical nature of some older museums, the emphasis on the power relationships established by other practices or traditions, the devolution of decisions to "experts" whose views are accepted rather than challenged, and the reluctance to address issues including issues of quality are both an abrogation of responsibility on the part of executives and not typical of successful organizations and their leadership.

Cohesion is advanced when the values of the organization are matched by those of all the people involved (Newlands, 1983). A convergence of people, power, and structure focusing on specific tasks, objectives giving play to sensitivity, creativity, and independence of thought, and relying on expert power rather than personal power or authority provides the climate that characterizes museums as learning organizations—for staff and visitors alike.

ENDNOTES

1. Preliminary results of this by Seig and Blankman-Hetrick (2006) show that visitors are staying longer at each area, a more dialogic style of interpretation is being used, and family conversations show richer and deeper discussion about what is being experienced.
2. Institute of Museum and Library Services.

12

Fostering Effective Free-Choice Learning Institutions: Integrating Theory, Research, Practice, and Policymaking

Jeffrey H. Patchen and Anne Grimes Rand

WHY INTEGRATE THEORY, RESEARCH, PRACTICE, AND POLICYMAKING?

Integrating theory, research, practice, and policymaking is not only important for the sustained improvement of a visitor's free-choice learning experience, it is essential for the development and communication of internal and external institutional policy, and ultimately, for the successful attainment of a cultural institution's mission, be it a museum, library, or other free-choice learning institution. The purpose of this chapter is (1) to make the case for why such integration is critical to the future success of free-choice learning institutions; (2) to identify the benefits of committing human and financial resources from within and outside the museum toward this effort; (3) to discuss the important role of internal policymaking to support the institutionalization of this integration; (4) to reinforce the importance of external policymaking and the role that each institution can play in furthering knowledge within the field of free-choice learning specifically and lifelong learning in general; and (5) to suggest specific steps that free-choice learning institutions can take to begin such an effort. This chapter has direct implications for key decision makers in free-choice learning organizations, such as directors, senior management, educators, board members, and funders interested in supporting healthy learning organizations. We present two perspectives: one at a large children's museum, and the second, a smaller institution also strongly committed to free-choice family learning.

In the ideal free-choice learning institution, professional staff and volunteers work diligently to meet the real and perceived learning needs of a wide variety of audiences. Those responsible for creating meaningful learning experiences within the institution continually seek, appropriate, integrate, challenge, and contribute to a body of research that will ultimately strengthen both the institution and the field of free-choice learning. These individuals—"chief worriers" in support of free-choice learning, if you will—serve as internal policymakers, seeking ways to effectively communicate an approach to learning that creates meaning for their audiences as well as for the larger group of staff and volunteers within their institutions charged with providing dynamic physical and virtual learning spaces, places, and programs.

Trustees and funders of nonprofit free-choice learning organizations also seek to understand just how the institution under their care and guidance can continue to improve, grow, and meet the needs of an ever-changing world in terms that make meaningful sense to them. Typical trustees within many institutions include businesspersons, lawyers, accountants, and community leaders. In their world, feedback loops and systems delivering data assessing the impact of products or services are regular fare. Building on their understanding of the importance of feedback in their fields can help these decision makers understand that creating systems for collecting and sharing data about free-choice learning, as well as mechanisms for using that data to inform and improve the visitor experience (aka mission return), has an important and central institutional role, alongside attendance and proper fiscal management. It is also important for trustees to understand that essential feedback includes the difference the institution makes in the *lives* of its visitors, as well as the number of visitors touched by the products and services created by the institution.

COMMITTING RESOURCES FOR INSTITUTIONAL ENGAGEMENT

An important goal of internal policymaking in a nonprofit free-choice learning institution such as a library or museum is to strengthen the work of professional and volunteer staff engaged in meeting the learning needs of visitors. Utilizing a theory-to-practice research model with implications for internal and external policymaking is one effective way to strengthen institutional practice. This engagement can take a number of forms along a continuum of involvement, commitment, and institutionalization. In some institutions, the director of education or a team of educators may initiate and nurture the engagement. In others, a team of educators and colleagues, in partnership with senior management and board representatives, may begin and continue the work. Some ways to utilize such a model in the planning phases include:

- Identify, document, and share core values for the education component of the institution's work that includes free-choice learning to guide the institution's approach and commitment to free-choice learning precepts.
- Identify, document, and share specific characteristics of what makes for an effective free-choice learning physical space or virtual space within and across the institution.
- Identify the importance to the institution of collecting, recording, and using data about and from visitors to improve its work.
- Identify the importance of informing and providing training for staff and any volunteers responsible for designing or delivering free-choice learning to visitors.
- Identify the importance of collecting data from visitors about their learning experiences on site and after returning home and analyzing it to inform and make adjustments to the institution's annual work plan and strategic plan.

Each level of engagement requires human and financial resources to implement the model successfully, particularly if continual training and assessment is taken on. The greater the extent of commitment and implementation, the more important it will be to identify what level of engagement is being pursued, who will lead and execute the engagement, and how results will be used and shared. Two specific examples—one at a large children's museum, and the second at a smaller institution—serve as illustrations.

Institutionalizing Free-Choice Learning: The Family Learning Initiative at The Children's Museum of Indianapolis

At The Children's Museum (TCM) of Indianapolis, the museum is engaged in a free-choice/family learning initiative that began in 2002 led by the museum's president and CEO and vice president for education and experience development, in collaboration with the Institute for Learning Innovation. This collaboration included an initial assessment of the museum's family learning approach in a specific permanent exhibit gallery, an NSF-funded temporary exhibition, two youth programs, and the initial visitor orientation experience.

Following these initial assessments, the museum continued to collaborate with the Institute and began an initiative to define what free-choice/family learning meant for the museum's education and experience development activity with a focus on the development of a new permanent science exhibition. By early 2004, a new definition of family learning emerged for the museum that was widely shared with staff, volunteers, and trustees: Family learning includes the products and processes of social interaction, collaboration, and sharing among members of a multigenerational group (two or more people but at least one adult and one child who have an ongoing relationship) across the lifespan of the family.

Front-end, formative, and summative assessment strategies were used to create and assess the new science exhibition, and a detailed professional development family learning curriculum was collaboratively developed with TCM and implemented by the Institute to immerse full- and part-time staff, as well as volunteers, who would work in the new exhibition in the principles of family learning. These family learning leaders continue to mentor and support their staff in efforts to facilitate family learning. An introductory document was created for new employees, trustees, and those interested in the museum's approach to family learning.

Because the museum and its trustees also want to know how each gallery learning experience relates to the museum's new definition of family learning, museum staff collaborated with the Institute to create an annual assessment plan and tools called the Family Learning Assessment System (FLAS). This tool assesses the museum's family learning work across all gallery learning environments annually. An important part of FLAS is a separate assessment tool to examine in depth a single gallery each year. In 2004, professional staff from the museum participated in presentations at national conferences, sharing the results of their work. FLAS was fully implemented in 2005. As the museum continues its commitment to integrating theory-to-practice research and policymaking in 2007, the museum is focusing on new assessment tools for interpretation in family learning settings and the development of criteria for defining a virtual family learning environment. Because of its size, The Children's Museum of Indianapolis has been able to engage in a long-term, comprehensive collaboration with an organization like the Institute for Learning Innovation, allowing these efforts to be successfully sustained by building internal capacity to adjust, implement, refine, and improve practice (for more details, see Dierking, Andersen, Ellenbogen, Donnelly, Luke, & Cunningham, 2005).

Laughing and Learning Together: From Theory and Research to Practice and Policy at the USS Constitution Museum

In smaller organizations, the path to sustainability can be difficult since continued dependence on "full-service" outside consultants is too costly. Thus it is critical to develop more organic and focused ways to institutionalize these processes. The USS Constitution Museum is taking such an approach by focusing its efforts on integrating family learning into one exhibition space.

Located on Boston's Freedom Trail, the USS Constitution Museum welcomes a large family audience (250,000 annually) that has traveled to Boston "to do history." Right across the dock from "Old Ironsides," the famed U.S. Navy warship undefeated in the War of 1812, the museum ensures that the stories of USS *Constitution* and those who shaped her history are never forgotten. But how to present stories from 200 years ago in ways that are meaningful and relevant to today's visitors? The museum has always blended hands-on learning with history, but in 2004 they decided to push further and create an exhibition with a primary purpose of engaging a family audience. This sounds straightforward; however, they said that family learning techniques would shape the exhibition first, rather than making the primary goal conveying as much history content as possible, an unusual approach for a history museum. They had already developed exhibition themes with a panel of scholars funded by a National Endowment for the Humanities (NEH) planning grant, but in designing the exhibition experience stated that family learning was their primary goal.

This vision led them to family learning experts Minda Borun at the Franklin Institute Science Museum and Dr. Lynn Dierking at the Institute for Learning Innovation. With input from Dierking and Borun they applied for an Institute of Museum and Library Services (IMLS) National Leadership Grant to explore effective techniques for engaging a family audience in an exhibition where there are no facilitators present, and built in collaboration time for Borun and the Institute. In their research prior to submitting the grant, they looked for models of successful *unfacilitated,* yet *interactive,* history exhibitions and found few examples. However, science and children's museums did seem to offer models for exhibitions that more effectively integrated family learning and the presentation of content. Could they apply these models to a history exhibition? Could they make history more personally relevant, presenting narratives in ways that foster personal connections between visitors and stories of the past? Could they identify successful, low-cost techniques to engage families in learning conversations that could then be shared with other small history museums?

Fortunately they received funding from the IMLS and have created a prototype exhibition called *A Sailor's Life for Me?* to explore these questions. Their guiding principles include applying research and successful techniques used elsewhere, such as Borun's PISEC criteria (Borun et al., 1998) that identifies seven characteristics of family friendly exhibits. They also have developed a matrix of learning styles, similar to models from children's museums in Indianapolis and Boston, to ensure that their exhibits engage people with different learning styles. They began with a thematic social history framework that emphasizes the personal experience of sailors aboard the *Constitution*, rather than a traditional chronological approach to a history exhibition. This allows more opportunities to foster personal connections and conversations, while freeing them from specific dates, times, and places.

The exhibition engages visitors in a cruise aboard USS *Constitution* during the War of 1812 with many opportunities for hands-on interaction, conversation, and fun. Visitors explore the daily life of a sailor as they sleep swinging in a hammock, eat their mess picnic style on the deck, climb aloft to furl a sail, and scrub the deck on their knees. The tone of the exhibition changes from fun to fear when it is time for battle, since a preliminary study of family visitors' perceptions encouraged them to present the realities of battle—not to sugarcoat it. This past summer, they built a mock-up of an object theater that integrates the actual words of 1812 sailors with images and artifacts to present the human experience of battle in order to test its use and effectiveness. The goal is to create an emotional experience, so the program presents the dreadful anticipation before the fight, the organized chaos of battle, and the bloody aftermath. The emphasis is on the personal experience of the sailors in this audiovisual presentation and throughout the exhibition.

Exhibition text is written in the first person, limited to 50 words, and presented on full-scale photo cutouts representing actual sailors who served aboard the *Constitution* in 1812. Throughout this experience, staff have firmly stated that encouraging family learning does not mean "dumbing down" the content; instead, it means presenting it in engaging ways. They actively seek opportunities to convey their themes through physical experiences, sound, touch, graphic charts, images, objects, or reproductions. Text panels are not the starting point for exhibition content; they are one tool among many. They think carefully about layering information and invite visitors to experience aspects of a sailor's life. Concepts such as teamwork aboard ship are much more memorable when experienced, rather than read in a text panel. They believe there are many possibilities for drawing on universal themes and making history come to life, rather than presenting it in dry and dusty ways.

Extensive audience research, led by Borun, who developed the methodology and trained *Constitution* staff and volunteers to collect data, has been integrated throughout the exhibition development process and has led to many modifications to make exhibit elements more successful. It also has helped the exhibition development team to recognize the value of including the audience at the exhibit development table early on. Testing concepts early and often helps to refine the concept, meet visitor's needs, and ultimately saves money—plus, it is far less likely that ineffective exhibit elements will be created—a win-win-win situation all round (for more information, see www.familylearningforum.org; Dierking, Rand, Solvay, & Kiihne, 2006).

INTERNAL POLICYMAKING

As demonstrated in these two examples, when embraced by an institution's professional and volunteer staff, effective internal policymaking can institutionalize commitment, encourage habits of mind about free-choice learning, and ensure consistency and improved reliability of data collection from various audiences. Internal policymaking is also an important internal communication tool, a means of sharing the core values, beliefs, and means of assessment to which the institution holds itself accountable with its primary audiences.

There are many examples, formal and informal, of ways to create and implement internal policy. Most typically, policies are created by senior leadership, a director of education, or a team of those responsible for creating and nurturing free-choice learning, though many institutions offer opportunities for staff to share their input as policies are developed. Some ways of creating and implementing internal policy include, but are not limited to:

- After defining what free-choice/family learning means for the institution, examine the organization's core values, mission, and vision to determine whether it includes some reference and/or commitment to free-choice learning.
- If the institution's core values, mission, and vision do not include some reference and/or commitment to free-choice learning, adopt or create and then share a set of core values that includes such a commitment.
- Make sure the institution's annual and/or strategic plan references specific objectives, strategies, and/or tactics that address the institution's commitment to free-choice/family learning.
- Ensure that new employee, volunteer, and board orientation includes an overview of free-choice/family learning.
- Create and implement a formal professional development program for staff and volunteers to ensure the institution's approach to free-choice learning is known and shared throughout the institution.
- Create an annual assessment tool or procedure that contributes to the collection of longitudinal data to support the institution's mission, vision, and core values.
- Document and share results among all levels of staff and volunteers inside the organization and provide opportunities to discuss the implications of the findings and ways practice can be improved.
- Apply these findings to all new free-choice learning projects, programs, and collaborations.
- Continually seek research and implementation partners to assist with the collection and analysis of free-choice/family learning data.

- Ensure regular reports are shared with trustees and funders to keep them apprised of progress toward improving free-choice/family learning with key audiences.

As with all good policymaking, regular review and revision is important. As data are collected and results affect annual work plans and the institution's strategic plan, the institution's leadership should reexamine, revise, and update policy-related documents (e.g., department core values, orientation overview presentations, etc.). Trustees should have the opportunity to be included in such revisions. If there is an education or experience development trustee committee or working group dedicated to working with staff on issues in this area, their involvement is a good way to ensure continued support and buy-in at the institution's governance level.

By institutionalizing these efforts internally, such processes become an established part of the way the organization "does its business." For example, the family learning project at the USS Constitution Museum is led by the Visitor Experience Team, which includes the deputy director, director of exhibits, curator, and director of education/interpretation, with the support of the executive director. This approach to exhibition and program development is permeating this particular museum. In fact, a culture of experimentation and positive risk has developed as an outgrowth of the project: "Let's try it and see what our visitors think" is heard far more within the institution than ever before. This approach has brought renewed enthusiasm to the staff as they actively seek audience feedback, then modify offerings accordingly. Embracing theory and research as practice is reinvigorating the museum. The efforts are also influencing policy; they are at the heart of the museum's strategic plan, with all departments involved in some aspect of the effort. Embracing free-choice learning and building exhibitions to engage families in learning conversations helps to keep the museum relevant to its current audience, while building constituents for the future.

EXTERNAL POLICYMAKING

External policymakers such as funders, large formal educational institutions, and community leaders can play an important function in supporting institutions that embrace and engage in the theory-to-practice research model. External policymakers can create policies that directly affect the nonprofit free-choice/family learning institution. For example, state education agencies exert tremendous influence on what is taught in K–12 schools, thus encouraging or limiting access to free-choice/family learning institutions through activities such as field trips. Funders, both private and governmental, also exert great force on the ability of nonprofits to secure financial resources to implement programs that have a direct impact on visitors. In the 1980s and 1990s, the J. Paul Getty Trust led a 15-year movement to improve the quality and quantity of arts education in the nation's schools through regional consortia of museums, public schools, and universities. Funding these consortia at an unprecedented level for an extensive period of time, Getty greatly influenced the theory and practice of visual art education in the United States. In the Getty model, not only did participating consortia implement theory, they ultimately contributed to a vast network of research-based institutions committed to improving art education for K–12 students and their teachers.

It is clear among many private foundations that accountability among its grant/program recipients is key to their own trustees and those responsible for guiding the foundation's work. By participating in a theory-to-practice research approach, nonprofit free-choice/family learning institutions demonstrate their commitment not only to core learning principles but also their greater commitment to continuous measurement and improvement to determine whether they are accomplishing their goals.

Engaging in external policymaking about the nonprofit free-choice/family learning institution's theory-to-practice research approach can take many forms including, but not limited to:

1. Using findings from your studies to *inform, seek, apply for, and/or secure funding* from foundations and/or government agencies that embrace free-choice/family learning.
2. Ensuring that the institution's public relations and marketing efforts are engaged in promoting the institution's commitment to free-choice/family learning to local, state, and regional policymakers so that they are aware of the institution's commitment to free-choice/family learning and meeting the broader learning needs of visitors.

3. Documenting and sharing results among professional colleagues at local, state, regional, national, and international conferences.
4. Collaborating with like-minded institutions and sharing information about the collaboration with funders and other external policymakers.
5. Securing the services of professional counsel to promote the institution's values, commitment, and need for resources to support its work.
6. Collaborating with professional assessment and/or evaluation experts *to design* and implement a customized plan for creating physical/virtual learning spaces and programs.

The benefits of external policymaking activity can take time to realize. The most obvious benefit is that such policymaking demonstrates a commitment to engage in continuous improvement, to research new ways of meeting the learning needs of audiences, and to collaborate with experts and other like-minded institutions.

TAKING FIRST STEPS

Falk and Dierking have posited there are eight key factors that influence learning (Falk & Dierking, 2000). Although they developed the model with visiting individuals and groups in mind, these eight factors also influence institutional learning and engagement as well. Integrating theory, research, practice, and policymaking necessarily involves learning at the individual, institutional, internal, and external policymaking levels. And taking first steps requires an understanding of the various factors in the model and thus is complex and challenging. The eight factors that influence the implementation of a theory-to-practice approach vis-à-vis institutional learning include:

1. *Motivation and expectations.* It is important to ensure that staff members are motivated and fully engaged in the approach and have clear expectations for their role in the theory-to-practice process. Integrating theory, research, practice, and policymaking also helps institutions understand why visitors engage with their institutions and

how institutions can improve their engagement through continuous improvement.

2. *Prior knowledge, interests, and beliefs.* No institution begins or enters into a theory-to-practice research approach, or internal/external policymaking, in a vacuum. Each institution brings to the model its own set of experiences with visitors, with staff, with trustees, and with funders.
3. *Choice and control.* Each institution needs to choose and control the level of commitment, involvement, allocation of resources, and internal and external policymaking it will be devoting to the effort.
4. *Within-group sociocultural mediation.* Just as families are affected by the physical and virtual learning places and programs that museums and libraries create, so too are institutions affected by how they craft their internal policymaking efforts and the means of communication among staff, volunteers, and trustees.
5. *Facilitated mediation by others.* Institutional engagement with outside experts, other professionals, and like-minded institutions can greatly facilitate implementation of the theory-to-practice research model. In addition, how interpreters, professional staff, and volunteers who engage visitors in free-choice learning environments facilitate learning greatly affects the visitor experience within the institution and after departure.
6. *Advance organizers and orientation.* Creating purposeful means of sharing information about the practical application of the theory-to-practice research model is important for a shared understanding among staff, volunteers, and trustees. Activities such as professional development and new employee and volunteer orientation are examples of how multiple means of communication can foster effective understanding.
7. *Design.* Just as the purposeful design of physical and virtual space and place greatly affects how visitors engage and learn, so too does the purposeful design of the institution's commitment to engaging in the theory-to-practice research model affect its success institutionally. Communicating plans, achievements, and improvements internally and externally further strengthens the institution's commitment to continuous improvement.

8. *Reinforcing events and experiences outside the museum.* Recognizing and sharing results of participating in the theory-to-practice model should have a positive impact on the institution, its key audiences, trustees, funders, and other external policymakers.

Integrating theory, research, practice, and policymaking is an important way for an institution to document its own learning, as well as the learning of its visitors within and external to the institution itself. Ultimately, it is also essential to the successful attainment of a cultural institution's mission, be it a museum, library, or other free-choice learning institution. At the core of the best free-choice learning institutions is a commitment to engaging the public in lifelong learning, regardless of circumstance. This chapter has attempted to address the importance of utilizing a theory-to-practice research approach and how effective internal and external policymaking can benefit the institution's commitment to continuous improvement. Ultimately, the success of such engagement and implementation will depend on those key leaders and professionals who themselves are committed to lifelong free-choice learning. May the journey of integrating theory, research, practice, and policymaking be a rich learning experience for you and your institution!

13

Meaningful Collaboration

Beverly Sheppard

> No one should kid themselves, . . . collaboration is not fluffy work. It is hard, frustrating, and unremittingly real, but it's worthwhile and absolutely essential in this new age.
>
> —Caroline Marshall (2002)

Have you ever watched children decide to play a new game together? The process begins with great enthusiasm and energy. One or two in the group establish the basic rules. Minor disagreements arise but are quickly settled. The players are eager to begin. "This sounds like fun," they all agree, and the game gets underway. Not long into the process, however, one or more of the children begin to question the rules. Soon, someone exclaims, "That's not fair!" or "You always get to be the leader" or "You changed the rules." What started with eager anticipation often breaks down into disgruntlement, frustration, and at least one player walking away.

Imagine, however, if the game took hold, if new skills were learned, new players added, compromise emerged, and a new form of recreation became a part of a town's summer rituals! Everyone would gain! Although most collaborations or partnerships are rarely so successful, some do have the potential for being truly transformative and providing rich public value. This chapter addresses such collaboration—partnerships that build and transform, enrich our institutional capacities, and deepen our public service.

From the outset, this chapter acknowledges that meaningful collaboration is very difficult. Such efforts require strong institutional commitment at all levels—sufficient time and resources, artful communication, and a clarity of vision. The work can be arduous, the risks great, and the effort far beyond the original conception. Yet the encouragement to cooperate and collaborate, to network and partner, is central to the language of today's museums, their funders, and their publics.

As our world becomes increasingly complex and diverse, needed skills and resources may no longer reside in a single institution. They most likely require institutional sharing, and even more essential, the finely honed art of listening and responding to one's audience—all essential modes of practice in the 21st century. Ultimately, collaboration is not just about the joint delivery of a product, it is about sharing and shaping an essential experience in concert with the very community and audience we wish to serve.

CONDITIONS FOR COLLABORATION

Collaboration exists on many levels. It can be a simple cooperative project between two or more institutions or it can become a deep and systemic partnership, fusing thinking, language, and formats, and ultimately creating something totally new. At any level, it requires a fresh way of thinking and working. It may respond to internal or external needs, focus on a single project, or evolve into a long-term relationship. Two publications, *Museums for a New Century* (1984) and *Museum as Catalyst for Interdisciplinary Collaboration* (2002), define changing approaches to collaboration.

Museums for a New Century recognized many types of collaboration, noting that most grow out of an internal, institutional need—offering some kind of economic utility. One exception is the long history of museum and school partnerships, perhaps the oldest and most successful form of educational collaboration in the world of museums. These partnerships represent the kind of transformative collaboration that has changed institutions and developed an array of new skills and services. Just think of the number of museums with education wings, large school departments, teacher workshops, and an increasing number of experiments to bring the two institutions together through technology.

While the earlier publication identified the purpose of most partnerships as meeting an *internal* institutional need, the second, more recent publication noted *external* audience needs as the driving force, indicating the sea change happening in the field and society. *Museum as Catalyst for Interdisciplinary Collaboration* summarized a series of conversations hosted by the Museum Loan Network. These conversations probed the specific aspects of collaboration that occurred between museums and other kinds of institutions. These conversations explored collaboration as both a creative process and a transformative one, suggesting that the need for many different voices and perspectives in our work is so great that it can change our practices and outlooks in substantive and permanent ways. The group cited the following as drivers for collaboration:

- The ability and need to engage new audiences
- The need to expand and challenge individual perspectives
- The development of new understandings and meanings
- The discovery of new ways to operate
- The ability to accomplish together what cannot be done alone
- The extension of unprecedented public access to resources

Inherent in each of these drivers is an essential learning experience. To expand, develop, discover, and extend are all traits of learning at the heart of the collaborative journey. They should be recognized at the start of any such venture as ultimate and meaningful goals. Success in any of these areas may profoundly change an institution and open it to entirely new ways of working.

The imperative to think and act collaboratively is among the core values of many of our institutions. If we believe that museums and other free-choice learning institutions are central to developing and sustaining social, cultural, educational, and economic well-being, we are already on the alert for such collaborative opportunities. If we believe that our mission includes meeting the needs of actual and potential visitors, we are already seeking meaningful strategies to do so. If we are clear about out assets and strengths and honest about our weaknesses, we are already looking for partners to complement our work. The conditions for collaboration begin, therefore, with a clear understanding of our mission, a desire to improve our effectiveness, a sense of our strengths, humility about our weaknesses, and a desire to learn and grow as institutions and individuals.

THE COLLABORATIVE PROCESS

Successful collaboration is never a given, despite our high-minded intentions. A failed collaboration can be very dispiriting and disruptive. An initial acceptance of risk, therefore, is vital to all parties. What follows is the amalgamated advice of many organizations and meetings that have sought to define how successful partnerships are established and sustained.

"Take a partner to lunch" was one staff member's suggestion as a way to begin. Such an invitation acknowledges that the up-front development of a partnership requires sensitivity and time to explore possibilities. Transparency and honesty are also essential. All players must be willing to use the table of conversation as a platform for addressing differences, limitations, and fears. Taking the time to build relationships allows partners to get to know one another's staffs, programs, facilities, and audiences as thoroughly as possible and begins the process of building trust, which will anchor the project when the inevitable hard times arrive. The Museum Loan Network conversations suggested beginning with a "tender" set of expectations—a lovely acknowledgment of the precious and sensitive beginnings of working together.

It is also essential to begin with a clear sense of mission. Mission identification should include an explicit articulation of why each partner is interested in collaboration. Partners need frank and open discussion about how their priorities, staff expertise, and resources will align to carry the project forward. A project that is peripheral to any one institution's core values and mission will never receive the prioritizations of time and funding that are critical to success.

Full support must come from deep within each museum or organization. The internal culture of each partner needs to be analyzed for potential barriers as well as for meaningful support. Individual staff allegiance to a project is not enough if all of the potential players—and decision makers—within the organization are not equally on board.

Cultural differences must also be acknowledged. Such differences are frequently expressed in the interpretation and use of language. When Wisconsin Public Broadcasting first began work with the Wisconsin Historical Society, for example, they quickly discovered that the meaning of "right now" had very different interpretations—from the immediacy of broadcasting to the more deliberative approach of museum decision making. Staffing and departmental structures, operations, and titles may also be quite different, leading to frustration in finding and sustaining the best internal partners when a project is underway. As all organizations do, museums have their own lingo, terminology, and ways of doing things, from job titles to professional practices. Communication depends on clarifying this language. Even more sensitive are areas of cultural and ethnic meanings, where the differences in language need to be carefully examined and discussed. Not understanding and openly discussing language use can lead to unintended but deep offense.

Readiness is another critical attribute in selecting partners. Each institution must have both a strong sense of its own assets and a willingness to grow and change. This is especially important in the area of shared authority. If ultimate authority is seen as the province of one of the players, then the collaboration is doomed to failure. A significant value of working together is the expansion of shared authority and the subsequent learning that occurs. These outcomes can be achieved only if an institution is open to them.

Even the best collaboration requires strong leadership. Every project needs internal institutional champions who can keep the projects going and drivers who can bring together the talents and skills of a diverse team. Deborah Kmetz, local history specialist with the State Historical Society of Wisconsin, describes the skills of such leadership as the ability to listen and the desire to learn the strengths that have helped shape the partnership, creating a supportive environment in which people can both contribute and compromise (Benton Foundation, 2001). She describes the attributes of a strong collaborative leader in the following words:

> You need people who are not egotistical. They can't have personal agendas. You need people that, when it's all said and done, the content is what moves them. Someone else's story ought to rivet them as much as the story they brought to the table. You need selfless people, but people who know where to draw the line. (Benton Foundation, 2001)

Most projects require a designated project manager to help shape the creative space, move the project forward, and maintain equity and balance. The project manager keeps the project on track, encourages full participation, and brings together the liaisons from each organization. A good manager finds ways to celebrate the combined wisdom of the whole, while drawing out the work required to ensure that the project remains on schedule.

Clearly good communication is essential. Finding ways to suspend judgment and reactive responses is hard work. It requires courage to place conflicts on the table and work toward resolution before they sabotage the collaboration. Each organization should periodically renew its commitment to honest communication. Process, management, and deadline problems will emerge in the best collaborations and need to be resolved in timely and transparent ways.

Perhaps the most important aspect of the collaboration among free-choice learning institutions is that it be "learner centered." Collaboration links the institutional partners to the public in new ways and respects the learner as a critical part of the process.

Essential to being part of a learner-centered collaboration is building opportunities for the learners themselves to participate in our work and contribute to its breadth and effectiveness as part of the process. When the motivation for collaboration is, indeed, public need, then bringing our audiences into the planning process is essential. We will be successful not when we say so, but when our public says so, and their voices must be heard and respected from the beginning. In the same way that we address the sharing of authority among institutions, opening ourselves to the authorities of our varied audiences can enrich and fulfill us, making our institutions true learning organizations inside and out.

INVITING MANY VOICES

Nowhere is the imperative to share authority more critical than in the potential partnerships that exist between museums and their communities. For far too long, museums have been on the sidelines of their communities, regarded as places rather than as resources. Community-based collaboration opens the door to expanding the role of museums as community leaders, catalysts for change, and responsive educators. However, this approach requires a new mind-set, one willing to let go of authority in exchange for open listening.

The need for building external partnerships with varied audiences has special urgency in our increasingly diverse world. Here again, our long-held notions of authority are increasingly under fire. An object in the collection of an art museum, for example, that comes from another culture takes on an increasingly complex meaning as those cultural representatives address its role in their own lives. There can be enormous distance between the meanings assigned by the professional and by the perceived "novice," who is often the "keeper" of family or community meaning. Every effort must be made to balance institutional authority with personal and cultural understandings.

Handing over the interpretive process to others, whose interpretation is usually personal or rooted in tradition, requires a willingness to accept that there are other criteria than scholarly research for knowing and understanding. If museums are to be effective collaborators with our diverse audiences, then we must develop strategies that respect the truth of another's viewpoint. We must be willing to solicit and accept multiple voices and points of view, to encourage respectful debate, and to create a more fluent and realistic portrait of history and community.

A bold example of expanding voice and interpretation took place at the Baltimore Museum of Art in spring 2000 through a multilayered project showcasing the work of local African American artist Joyce Scott in an exhibition and multilayered project called *'Kickin' It with the Old Masters.* Scott is renowned for striking fiber pieces and beaded sculptures that offer biting social commentaries on such issues as racism, violence, sexism, and stereotypes. Imagine entering the formal entrance court of the Baltimore Museum of Art and seeing a lynched body of a black man, fashioned from glass beads, hanging above the powerful bronze figure of Rodin's *Thinker*! Such provocative juxtaposition was an essential part of the landmark show (Wallace Foundation, n.d.).

The stunning collaboration between the museum and Scott sought to engage new audiences and challenge traditional thinking. Scott's work appeared throughout the museum, accompanied by programming and engaging interactive and hands-on experiences. It could not have occurred without inviting Scott to use the language of her work freely and fully. Not only did the collaborative process entice new audiences, many of whom had never felt at home in the museum, it also aggressively challenged traditional audiences to reconsider the role of art in contemporary life. Both the artist and the museum's director Doreen Bolger recognized that the venture took them down new roads, but began with a "congress of minds" (Wallace Foundation, n.d.). Bolger believes that the exhibition changed how people think and how they now think about the museum: "Basically the museum is a different place because we did the exhibition. . . . I think it's changed how the public perceives the museum. That's a long-term legacy" (Wallace Foundation, n.d.).

Clearly, collaborating with our communities offers new challenges. Our metaphors for re-creating experience become far less linear. Our organizational patterns for new exhibitions, programs, and publications need to parallel the complexities of life—they are quilts, tapestries, and mazes rather than timelines and static displays. They are fresh opportunities to allow many ideas and minds to review the evidence of past and present. They also must create the conditions for inviting others to participate in our decisions and the creation of our exhibitions and our programs. Museums are challenged to find and welcome voices within their audiences to share in the collaborative processes of seeking mutual understanding and respect.

The Wing Luke Asian Museum in Seattle, which is founded on the principle of community partnership, uses collaboration to transform the traditional meaning of museums. As projects are developed, community members become part of the museum's staff. They add the authentic stories, perspectives, and concerns of the community to the traditional skills of museum-trained staff. Ron Chew, the museum's director, asserts that the staff must have "the skill and stamina to activate relationships" (2003). He urges a real transfer of power as the essence of transformation, not a symbolic advisory board. These same principles apply to building partnerships between communities and museums, finding a balance of authority, skill, and equitable ownership.

If museums are to be essential to the circle of community, they must be conscious of their perspectives, willing to engage the community in reaching in, bringing language, content, and substance from the "outside world" with them. This shift toward greater inclusion and community engagement is fraught with serious challenges to traditional museum practice, especially in the areas of scholarship, internal authority, exhibition development, and programmatic activity. Self-examination and readiness are critical to success.

Museums engaged in new collaborations with communities also must be realistic about the resistance they may experience. The disconnect between museums and many of the new audiences they wish to reach is vast. Whole segments of American society have found no records of their existence in museums or only records that have been misrepresented or fragmented. Museums must earnestly work to build trust and invite honest dialogue. They must be open to new ideas about community collecting and collections, and must peel back the layers of their own practices, inviting the public to understand the processes and patterns of museums at work so that they can find a meaningful place within them.

If museums are to be useful players in their communities, they will need to be more visible and more actively involved in community issues. As long as they remain on the periphery of their communities, they will not achieve recognition as leaders or even as possible partners. The American Association of Museums recently hosted museum/community dialogues in six cities—conversations often filled with surprises and insights. Gradually, the civic participants began to see museums as more than "lovely" places. They began to identify the assets that museums bring to public discourse. They noted that museums offer expertise and skill sets, collections, gathering spaces, and a strong sense of neutral territory in which to confront differences. The museum as a safe place, a community front porch, offered a fresh take on museums as facilitators and team players, not just as authorities and repositories. Such recognition can be the foundation for highly effective museum and community collaboration. However, it requires being at the community table in as many conversations as possible.

COLLABORATION AND TECHNOLOGY

Museums have many communities beyond their surrounding neighborhoods. Technology facilitates communities of interest that reach far and wide—from our sister institutions across the globe to the enormous breadth of interests represented by our collections and expertise. Many of the most exciting and powerful collaborations at hand are those that build upon the great promise of technology.

Consider how closely the language and structure of collaboration, from linking to networking, aligns with the language and structure of current technologies. The image of a net with an infinite number of intersecting nodes offers a powerful metaphor for collaboration. Technology provides many tools for building and sustaining new partnerships. For example, digital platforms can now assemble massive collections from museums, libraries, and archives, exceeding the limitations of physical space. Technology offers the promise of unprecedented accessibility, and it can foster and sustain extraordinary links between users and providers. The learners seated before their computers are the epitome of self-directed learners, driven by their own learning needs, personally selecting and assembling resources most suited to meeting those needs. Their process of learning will be highly individualized, limited only by the availability and usability of the resources online. And frequently those resources are simply waiting to be made accessible.

Even deep, transformative collaboration can be facilitated by technology. Consider, for example, the effectiveness of the Colorado Digitization Project in providing public access to museum collections across a state that often poses geographical challenges for the traveler with its vast distances and rural nature (Collaborative Digitization Program, n.d.). The Colorado Digitization Project, established in 1998, represents one of the most successful and influential online collaborations, one that has continued to grow and evolve and serve as a replicable model. The project's original purpose was to explore how to link the resources of many of Colorado's museums, libraries, and archives, bringing together disparate resources to capture and share the state's heritage as widely as possible. The project's organizers established unique opportunities to fund and provide digitization opportunities and provided mobile digitization laboratories to remote sites.

Today the project has grown to include ten additional Western states and its new name, the Collaborative Digitization Program, reflects its growth. It provides a Digital Toolbox of resources for converting analog materials to digital formats. It operates workshops on many related topics and hosts a Teacher Toolbox of lesson plans and links to primary sources designed specifically for classroom use (Collaborative Digitization Program, n.d.). Central to the success of this project is a commitment to collaboration. Working groups, representing the partners, have established the frameworks for each aspect of its development. New groups form when the project discovers needs for additional expertise, such as standards for digital audio. The far reach and effectiveness of the project is also the basis for ongoing funding, as funders can readily see the impact of this very well-managed, ongoing effort.

This project demonstrates how collaboration can help "connect the dots" for a public seeking to extend its free-choice learning experiences. As American and world societies increasingly become "learning societies," the need for such free-choice learning infrastructures becomes increasingly important. Our institutions offer numerous opportunities for free-choice learners to fulfill their lifelong learning needs and special interests—but access to and information about these institutions and programs are often difficult to find and highly disconnected. A true learning society should provide easy to find, integrated, systematic, and equitable access to learning resources. One of the most significant goals of collaboration should be to find and experiment with ways to build and sustain such learning networks.

An encouraging sign that recognition of the free-choice learning society is emerging can be seen in the number of public and private organizations that are encouraging learning partnerships. For example, in its 2000 publication *Connecting Communities*, the Benton Foundation argued in behalf of using new technologies to create community media platforms

> to form alliances which contain all of the principal interests and institutions that make up the community. They will range from state and local governments to small groups of caregivers and ethnic associations. Most importantly, they will include the whole network of learners and teachers that make up the educational fabric of the community. (Somerset-Ward, 2000)

Similarly, *Museums, Libraries and the 21st Century Learner*, a 1999 publication of the Institute of Museum and Library Services, stated:

> We must become a nation of learners—individuals, families and communities engaged in learning in our schools and colleges, libraries, museums, archives, workplaces, places of worship and our own living rooms. Our experiences may be real or virtual, hands-on or on-line, as we engage with resources found

throughout our communities or available through television, radio, the Internet or the integrated technologies of tomorrow. (Sheppard, 2000)

In response to the publication and a subsequent series of conferences and discussions, the Institute of Museum and Library Services and the Corporation for Public Broadcasting have created a joint funding opportunity to encourage collaboration across disciplines. The first year of funding includes projects in science literacy, community health, historic preservation, and help for drug and alcohol abuse. These innovative projects are dramatic examples of the power of building learning networks and collaboratives.

ADVOCACY AND ASSESSMENT

Collaboration offers many significant outcomes—not the least of which is an opportunity to create public awareness about the value of museums as learning institutions. Just as the potential impact of a project is enriched by bringing many strengths to the table, so is the strength of our institutional voice. One of the most discouraging concerns among museum professionals is the ongoing lack of public awareness about the breadth of what a museum can bring to community life. We are thought of (fondly) as wonderful places to visit. We are sought out during our blockbuster shows, our signature annual events, and most frequently on weekend days. Few of our visitors think about us as resources to tap into in ways similar to how they visit libraries—seeing us as resources for countless ideas and information for the gathering. Even our school visitors, the busloads that arrive on warm spring days, do not often consult with us as they develop new curriculum or seek new learning methodology. Museums are generally seen as positive places—though often a bit remote—but we have yet to be fully perceived as resource-rich leaders in a changing world of learning.

Strong, effective free-choice learning partnerships can move us closer to that elusive recognition. So much of what occurs during the course of collaboration is like building a new friendship. In the best of worlds, all will go well. We will invite and listen and hear one another's stories. We will match our interests and our strengths and find points of rapport. We will learn from one another—opening our own understandings to a world of new possibilities. And when we return home, we will most likely talk about our new friendship and share with others how it can grow and be part of our institutional practice. Eventually, as we learn to trust one another ever more, we will confide our dreams to one another and ask for feedback. We will shape collective dreams in this way and refine them as we go. We are likely to disagree occasionally and to argue, but if our friendship is to last, then our differences will be noted and worked through. And even though our combined goals are to reach out to others, we too will be enriched. If that sounds a bit self-serving, we can ponder the meaning of friendship and note that everyone gains. As our circles widen through such ongoing efforts, then gradually the recognition we seek will also come.

Not all collaborations will unfold so smoothly, however, There will be times when we find them too demanding and draining. They will consume far more time and energy than we first assumed, and we will be tempted to step away. Therefore, it is critical that we conduct ongoing and intermediate check-ups, that we set up agreed-upon points to assess where we are in terms of our initial goals and outcomes. In a collaboration, if one of us fails, so do all partners. Tune-ups are possible and, in fact, usually necessary. The commitment of resources always requires justification in a manager's eyes, and the very future of sustaining partnerships may be at stake if we fail to do so. If scarce resources are to be invested, then we must identify our evidence of success. Just as the topic of evaluation is increasingly on everyone's mind in the design of exhibitions and programs, it may be even more critical in the collaboration business.

The third Museum Loan Network conversation concluded that *project evaluation* should be concerned with the following areas:

- The *problem* the collaboration addressed. Was it the right one? Was it defined correctly, narrowed or expanded in the most meaningful ways?
- The *vision* for it. Were the goals the appropriate ones? Did they end up matching the needs most closely?
- The *strategy*. Were the best means employed? Did they draw on each institution's strengths?

- The *implementation*. Was the timing right? The plan on target? Was it complete? Of quality?
- The *impact*. Were the desired results achieved? How did they transform each partner as well as the public?

Other questions of a transformative collaboration also need to be asked. Were more talents and expertise available? Did resources increase? Did the content become richer? Is it clear that the project was better because it drew people together? What did each organization learn in the process? Were any efficiencies gained? Where are the opportunities to share the lessons learned?

Collaboration is a creative and imaginative process. It is a big-picture activity—one that allows us to dream of the possible and open our eyes to new perspectives, learning from the experiences, talents, and histories of others. The collaborative process dreams and discards, energizes and eliminates. It starts and stops, and participants may wonder if all of the effort is even worthwhile. Collaboration should be an "opening" process. If we are aware of our own growth, of our own professional worlds being transformed, and of our audiences being better served, then we have all of the incentive we need to keep moving forward.

IV

INVESTIGATING MUSEUM LEARNING IN THE NEXT TEN YEARS

14

Understanding the Long-Term Impacts of Museum Experiences

David Anderson, Martin Storksdieck, and Michael Spock

The value and importance of understanding the long-term impact of museums on visitors should not be underestimated. Such information enables museums to understand how to improve visitor experiences in museums, as well as the subsequent impact of those experiences, in a multiplicity of dimensions. These dimensions may include the enjoyment visitors feel, the kinds of things they learn, or the degree to which they develop understandings or appreciations of the messages museums communicate. Understanding how these dimensions of impact sustain, emerge, change, and diminish over time provides valuable information about how to improve museum experiences for visitors. The nature and quality of learning and enjoyment derived from a museum visit may shift significantly over time and the true impact from the museum visit may not actually occur during the visit, but afterwards, through subsequent experiences. If these experiences were caused or motivated by a museum visit, the true learning outcomes would only be fairly assessed if that follow-up by the museum visitor is taken into account. The long-term impact of museums should be considered not only at the level of the visitor but also at the level of the communities museums serve. Thus, understanding the long-term impact of museums enables a better understanding of how to serve and enrich communities, of which museums are a part.

This chapter explores what is already known about museums and their long-term impact on visitors, the complexities and challenges inherent in trying to study and understand long-term impacts, future research and methodological approaches that we can use to effectively assess the long-term impacts of museum experiences, and the implications of these efforts for practice.

THE CONCEPT OF IMPACT

Museums are visited for a multitude of reasons: for leisure and enjoyment, to spend quality time with family/children/friends, to experience something unusual, to take part in a culturally enriching activity, to "learn new things," and many more reasons, most of which can be summarized under "self-fulfillment."[1] Consequently, the impacts of museum visits, be they long or short term, reflect the visitors' agendas and span a broad range of experiences, from a life-altering experience to feeling slightly amused for a limited period of time. In addition, museums and similar out-of-school learning environments are used extensively for a variety of purposes by teachers who bring their students on field trips (Anderson, Kisiel, & Storksdieck, 2006; Griffin, 2004) and there is increasing pressure to demonstrate the short- and long-term benefits of these experiences.

Research in the visitor studies field has often focused on "learning" as an important outcome of a museum visit. "Learning" has very often been conceptualized as a cognitive process of reaffirming what is known, activating latent knowledge, or creating new knowledge at various levels of complexity. In addition, knowledge can be gained at different levels of complexity, from the simple awareness of things to declarative knowledge to highly complex conceptual understanding. Most museum scholars would contend that learning in and from a museum involves visitors who construct their own meaning and understanding—meaning and understanding that varies greatly depending upon the background, experience, and knowledge a visitor brings to the experience, the visitor's social group, and the sociocultural and physical context of the institution itself (e.g., Falk & Dierking, 2000; Hein, 1998). In this perspective, learning is best conceptualized from the visitors' perspectives and, if measured as long-term impact, needs to be based on and tied to the visitors' overall museum experiences. When learning from and in museums is broadly defined and based on the visitors' agendas, the subject of learning quickly expands well beyond the museum exhibits and programs, to include, among others, learning about museums as places for lifelong learning and that museums are places to learn about oneself and the people who accompany the visitor. Additionally, much of the research on impact and learning has considered the individual (visitor) as the unit of analysis, yet changes that result from museum experiences can be examined on different (larger) scales. For example, several studies have investigated the impact of museum experiences on family groups (Borun, Chambers, & Cleghorn, 1996; Briseno, 2005; Ellenbogen, 2002) or even an entire community (Falk, Storksdieck, & Dierking, in press; Jones & Stein, 2005). Aside from "learning," other outcomes that are relevant to museum visits are increased interest in a topic or subject, and subsequent higher motivation to learn about it, with resulting increase in attentiveness and exposure to subsequent reinforcing experiences. Given this complex set of potential museum visit outcomes, long-term impacts have to be conceptualized from the visitors' as well as from the museums' perspectives and need to be understood as broadly as the roles museums play in today's lifelong learning societies.

WHAT WE ALREADY KNOW

There are but a handful of studies that have investigated long-term impact arising from experience in museum and museum-like settings, and most consider the longitudinal impact only over relatively short time frames—weeks and months after the visitor experience (see Adelman, Falk, & James, 2000; Anderson, 1999; Anderson, Lucas, Ginns, & Dierking, 2000; Storksdieck, 2006; Storksdieck & Falk, 2003; Storksdieck, Ellenbogen, & Heimlich, 2005). These studies generally find that cognitive and affective changes that can be measured directly after a visit tend to decline even over a period of a few months unless the museum experience itself is followed up by subsequent reinforcing experiences or assumes a personal relevance in the biography of the visitor. Still, most learning indicators, while declining from the immediate postvisit measurement to one conducted weeks or months later, tend to remain higher than prior to the visit. In addition, there are several studies that shed light on the impact of visitors' memories of experience in such leisure-time settings (Anderson, 2003; Anderson & Piscitelli, 2002; Anderson, Piscitelli, Weier, Everett, & Tayler, 2002; Anderson & Shimizu,

2007; Bogner, 1998; Dettmann-Easler & Pease, 1999; Ellenbogen, Kessler, & Gillmartin, 2003; Falk & Dierking, 1990, 1992, 1997; Fivush, 1983; Hudson, 1983; McManus, 1993; Medved, Cupchik, & Oatley, 2004; Medved & Oatley, 2000; Stevenson, 1991; Wolins, Jensen, and Ulzheimer, 1992).

The various studies described above, and some others, indicate important aspects of long-term impacts from museum visits. These were summarized recently by a group of museum professionals as part of the *In Principle, In Practice* conference (Anderson, Bibas et al., 2006). The following presents an expanded list of this summary:

- Museums and other free-choice learning institutions are clearly capable of fostering memorable and even transformative experiences (e.g., Falk & Dierking, 1997; Spock, 2000a, 2000b). These experiences determine the value assigned by visitors to their visit and, in the aggregate, determine the value of museums for the communities they serve.
- Factors such as prior knowledge and interest, visitor agenda, the sociocultural identity of the visitor, or prior experiences affect how visitors engage with the museum environment, learn from the museum visit, and encode memory (e.g., Anderson, 2003; Ellenbogen, 2002; Falk, 2006.)
- Various aspects of the museum environment, including the quality of exhibits and the opportunity to make personal connections, play a strong role in attracting and engaging visitors and making the memory more salient, memorable, inspiring, and personally satisfying for visitors (e.g., Falk & Storksdieck, 2005; McManus, 1993; Stevenson, 1991). What visitors remember after more than a year is mostly contextual (Falk & Dierking, 1997). Apparently, visitors forget the details of the content, but remember almost everything else in ways that are tied to their biographies or personal agendas for the visit.
- Memories of visits to museums, like all memories, are not stable—they change over time. Long-term memories from a museum experience are shaped not only by the nature of the visit itself but also by the visitor's subsequent memories and experiences (e.g., Adelman, Falk, & James, 2000; Anderson, 1999; Bielick & Karns, 1998; Ellenbogen, 2002, 2003; Falk, Scott, Dierking, Rennie, & Cohen Jones, 2004; Medved, 1998).

- Memory is influenced by visitors' satisfaction, interest, and motivation, much as these factors are shaped by visit memories themselves. Affect and memory seem to feed each other (e.g., Anderson, 2003; Falk & Dierking, 1997; Spock, 2000b; Medved & Oatley, 2000).
- People have varying abilities to recall and reflect on experiences. While museum visitors may report on what is important to them at the time, they may expand on their recollections and reflections much later (e.g., Ellenbogen, 2003; Falk, Scott, Dierking, Rennie, & Cohen Jones, 2004; McManus, 1993; Spock, in press).
- Salient aspects of an experience often remain latent until a later time (e.g., Falk, 1988; McManus, 1993; Wolins et al., 1992).
- There is evidence to suggest that identity is a key factor in how museum experiences are processed, encoded into memory, and recalled (e.g., Anderson, 2003; Ellenbogen, 2002, 2003; Falk, 2006; Medved et al., 2004).
- Attitude is generally not influenced by brief museum visits and museum-like experiences (Storksdieck, 2006). Even if short-term attitude measures indicate change, without subsequent reinforcing experiences and follow-up engagement, attitude scores tend to fall back to baseline (Adelman et al., 2000; Ellenbogen et al., 2003). However, attitude change can be sustained if the original experience lasted for at least a day, if not longer (Bogner, 1998; Dettmann-Easler & Pease, 1999).
- Long-term learning from museums may depend on the measure used and prior knowledge and interest of visitors (Storksdieck & Falk, 2003; Storksdieck, Falk, Witgert, 2006). For some visitors, true learning might start only after the visit, while others may satisfy their quest for knowledge during the visit. Long-term learning seems to depend on initial learning, the type of learner, and the type of learning itself: shallow versus deep, conceptual versus declarative; long-term learning, at least for some visitors, might be connected to the ability to "digest" memories (McManus, 1993).
- Episodic memory is the explicit memory of certain events, such as the time, place, or emotions associated with the events (which affect how we memorize the event). Episodic memory is linked to semantic memory, the memory of facts and concepts. It seems that visitors create episodic memories with ease, recalling what they did and how

they felt during a museum visit (Stevenson, 1991); semantic memories, on the other hand, require subsequent reinforcing experiences or a strong personal connection to the topic or content to manifest themselves (Falk, 2006; Stevenson, 1991; Storksdieck, Falk, Witgert, 2006). What people remember easily seems to depend to a large degree on the initial agenda of and enacted identity during their visit: Family-oriented visitors remember more strongly who they were with while visitors with an interest in the objects may better remember the specific event (Falk, 2006; Falk & Dierking, 1990).

- Sharing experiences with others through conversations (Stevenson, 1991) or by expressing emotions of the visits such as enjoyment, curiosity, frustration, and anger (Medved & Oatley, 2000) helps shape and enforce memories and therefore the subjective impact of a museum visit. Visitors tend to rehearse memories of their museum experiences when they discuss and relive their visits with others. Visits that spur conversations are thus more likely to create sustained memories.
- Affective school field-trip memories have a strong influence on future visitation (Anderson & Piscitelli, 2002; Falk & Dierking, 1990). Positive memories of museum school field-trip visits are linked to mild novelty and unusual experiences (Hudson, 1983; Wolins et al., 1992), connections to a child's sociocultural and personal life (Anderson, Lucas, & Ginns, 2003), and even classroom connections (Gilbert & Priest, 1997; Jensen, 1994; Wolins et al., 1992), an indication that embeddedness and connectedness—whether into the personal or some other sphere—is most important for children.
- Very long-term memories sustained over years or decades seem to focus on the social context of a visit (Anderson, 2003; Falk & Dierking, 1990). The sociocultural identity of a visitor seems largely to determine what visitors perceive during an experience and what they ultimately recall afterwards (Anderson, 2003). Overall, three factors seem important in shaping vivid long-term episodic and autobiographical memories of visitor experiences, namely, rehearsal of the memories, emotional affect associated with the source experience, and the degree to which their planned agendas were fulfilled or frustrated (Anderson & Shimizu, 2007).

ASSESSING AND INTERPRETING LONG-TERM IMPACT

As the discussion of long-term impact studies illustrated, researchers have employed a variety of methodological approaches to assess and interpret the long-term impact of museum experiences. Common to most studies of impact was the use of salient memories, which were recorded either through face-to-face or telephone interviews. While some researchers used prompts like souvenirs as symbolic representations of meaning and impact, or photographs of the museum, many employed sequenced questionnaires to trigger memories. Fewer researchers measured long-term impacts psychometrically, either as "learning" or as attitude, interest, and behavior change. While measuring the long-term outcomes of museum experiences certainly poses methodological challenges, there is no broad agreement in the field as to the most important, comprehensive, appropriate, or acceptable outcome measures that should be used and that—ultimately—would guide the use of methods and methodologies (Dierking et al., 2002). As it is with all social science research, the research questions (or objectives) dictate the research design of the study and the methods (tools) employed (Gay & Airasian, 2006).

CHALLENGES AND ISSUES SURROUNDING ASSESSMENT AND INTERPRETING LONG-TERM IMPACT

The challenges in assessing the impact of visitors' experiences from museums are numerous. The challenges are also a function of the complex nature of human experiences, the tremendous variability in museum visitor experience, and also inherent to the chosen research methodologies employed to gain an understanding of the impact. The question remains: How can we assess the rich, complex, and highly personal nature of museum experiences, and specifically learning from and in museums, in valid and reliable ways? Dierking et al. (2002) concluded that the honest response to this question is "with great difficulty!" There are a plethora of factors that are threats to our understanding of impact, some which are obvious and others perhaps not so obvious. Some important limitations and realizations to consider in the understanding of long-term impact are discussed below.

1. The Museum Experience Reconstructs with Subsequent Life Experience

As described above, museum experiences are reconstructed with subsequent life experience following the visit. It is important to realize that impact from any experience over time is not static and that memories change: They can fade into oblivion, they can be rehearsed and thus sustained, or they can be redirected and changed (Falk et al., 2004). The impact of a museum experience can develop further, change, or diminish as a function of subsequent life experience. Both aspects, the dynamic memory and the change in the degree of impact over time, present an interesting dilemma—what aspects of impact can and should be attributed to the immediate experience, and when is the ideal time to measure impact? What aspects of an individual's memories of a museum experience are a result of the museum experience and what are attributable to the subsequent life experiences? This dilemma is compounded by the fact that subsequent reinforcing experiences can count as museum visit outcomes, as follow-up experiences, increased interest and attentiveness, or information-seeking behavior (Storksdieck, 2006). It is equally true that certain kinds of subsequent life experiences will enhance the impact of museum visits (Storksdieck & Falk, 2003)—but to generalize about impact is complex given the vast diversity of life experience that visitors encounter following their visits, and moreover, the set of subsequent life experiences that will be meaningful to the visitors and result in connections back to the museum experience. Thus, probing impact is never only a measure of the source experience alone, but rather a function of the impact of the source experience dynamically constructed and reconstructed with many other prior and subsequent life experiences since that experience (Falk & Dierking, 2000, 2002).

2. Visitor Reality Versus Objective Reality

It is well established that memories of personally relevant experiences construct and reconstruct longitudinally, and what visitors report may not be an entirely accurate account of how the experience that produced the memories actually occurred (Bruner, 1994; Freeman, 1993; Neisser & Fivush, 1994). However, it is important to realize that self-reported long-term memory ought to be considered the visitors' current reality of the recalled events, which may or may not be entirely representative of the original reality (Anderson & Shimizu, 2007; Schacter, 2001). Recent trends in memory research focusing on memory failure or distortion, such as false memory (e.g., Schacter, 2001), further justify the need to investigate visitors' long-term memories of experience at venues like museums.

3. Visitors' Abilities to Reflect on Experience Vary

Visitors' varying abilities to articulate their experiences pose a challenge in assessing long-term impact. Some visitors are highly reflective, while others, such as children, may not be able to reflect on their experiences, or may have difficulties verbalizing or expressing them. Thus, great care must be taken to ensure that the research methods that record memories or self-reported long-term impact do not create undue bias and that they are sensitive to the nature of the visitor. Large quantitative studies may avoid some of the biases by averaging effects over a sizable sample. However, qualitative studies with small sample sizes need to critically assess respondents' reflective abilities.

4. Who Are Visitors When They Enter the Museum Experience?

The notion that new knowledge (as one measure of impact) develops out of prior knowledge (past experience) is a fundamental tenet of constructivist theory, and has been demonstrated in both formal and informal learning environments (e.g., Anderson, 1999; Falk & Storksdieck, 2005). Applied to the broader range of outcomes for museum visits, visitors' prior interests, values, attitudes, knowledge, and motivations need to be factored into the museum visit outcomes—all key factors that may influence how visitors derive meaning and enjoyment in a museum setting. Yet it is almost impossible to measure in any acceptable time frame validly and reliably all the various psychographic factors that could influence a museum visit outcome, though first attempts are being made to develop short, closed-ended instruments that capture many of these factors (see Heimlich, Bronnenkant et al., 2005).

5. Instantaneous Measures Only Tell Part of the Story of Impact

It is important to realize that the impact of visitors' experiences throughout and at the conclusion of a museum visit are not static, but rather that visitors continue to construct and reconstruct their assessment and their memories as these become intertwined with subsequent experience in the days, weeks, months, and even years following the visit (Falk et al., 2004, 2006). What visitors value, and are hence able to articulate at a given instant, is contextual, and time is an important contextual variable. Thus, instantaneous measures of impact tell part of the impact story, but not the whole story. On the other hand, any time period studied later will also only tell part of an evolving story. This makes it impossible to declare an ideal time frame for conducting longitudinal impact studies. However, for the sake of comparability, it would be desirable if the research and evaluation community were to recommend particular time periods after the visit as sample periods for measuring long-term impact.

Assessing the impact of museum experiences should ideally include multiple "snapshots" of the impact over time—from initial experience to point(s) in time proceeding from the experience. Capturing longitudinal data is in itself challenging given the problems associated with contacting visitors after their visits and the resulting mortality (the loss of research participants in a study) that can be associated with certain kinds of long-term assessments. Post-hoc design studies such as Anderson (2003) are a useful way around these challenges given the caveats about visitor reality versus objective reality discussed earlier.

6. Assessment Interventions Have an Impact Themselves

The very act of probing visitors about their experiences has the potential to change their experiences and the overall impact of these experiences (Palys, 1997). This is an age-old problem in the social sciences—if one wishes to truly understand the impact of a museum experience on visitors, one cannot entirely discount the impact of the assessment procedures that visitors participate in on their overall museum experiences. The fact that we ask opinions or get visitors to reflect on their experiences actually changes their experiences in ways that would have ordinarily not occurred if we had not asked them. Typically, evaluators attempt to make the assessment interventions as naturalistic as possible. If the assessment methods have the characteristics of being casual, noncompulsory, engaging, rewarding, and not overly taxing for the visitor, then one might rightly argue that the experience was a natural and harmonious part of the museum experience. Some even argue that visitors, ultimately, follow their own agenda, and since interactions with museum evaluators are voluntary and thus become part of that agenda, they ought to have only a very limited impact on the museum experience.

7. Issues of Validity and Reliability—Probes and Interpretations

Face-to-face assessment methods, such as interviews, can be complicated by a plethora of factors that can influence visitors' responses about the impact of their experiences. Reactivity as social desirability or simple pleasing of the interviewer creates a need to interpret positive data carefully and to insert mechanisms that minimize reactivity in general. Perceived power inequality between interviewer and visitor can also have an impact on visitor responses. For example, young visitors may not feel entirely at ease in expressing their opinions to interviewers who are mostly older than they. It stands to reason that if visitors have needs or desires that do not include participation in assessments and evaluations, the usefulness and reliability of data about their experience will be compromised. While a variety of methods and procedures exist to minimize reactivity (listed in any good textbook on social science or education research), even experienced data collectors with well-designed and validated instruments will need to assess reactivity in the data analysis and interpretation phase of their research.

8. Sensitivity of Tools

The sensitivity of assessment tools and probes are critical for the quality of data that are generated, particularly in longitudinal studies. As previously discussed, there is a wide range of tools (instruments and protocols), methods (procedures), and methodologies (designs of research and evaluation studies) for gaining understandings of the impact of visitors' museum experiences. Sensitivity of tools, methods, and methodologies refers to the degree to which one can gain information, appreciate, discern, and hence understand the impact of the experience on the visit. Certain types of questions aimed at probing impact are more revealing than others. For example, asking *Did you enjoy your visit to the gallery; yes/no?* certainly provides information, but the leading nature of the question and the limited choices, and thus the overall lack of sensitivity will severely limit the validity of the results. Asking questions such as *What was it about your visit to the gallery that you enjoyed most/least?* provides a deeper level of sensitivity, and hence more appreciation about the nature of the impact, and asking enjoyment on a rating scale or semantic differential provides the choices that allow visitors to more appropriately express their perspectives.

The challenges we discuss here are but some of many that confront researchers who are concerned with understanding long-term impact. The challenges are not insurmountable, but it is necessary to acknowledge and address them in the development of research studies and research approaches that address long-term impact. Ultimately, all good research designs (both qualitative and quantitative, short term or long term) aim to minimize the threats to validity and reliability—flagging these issues serves as a useful reminder to guide the development of methodologically sound studies.

FUTURE RESEARCH AND METHODOLOGICAL APPROACHES

Finding valid and reliable ways to assess the impact of museum experiences beyond the actual visit is a challenge: there are few studies that have investigated long-term impact of museum or other leisure experiences and hence the literature on the topic is thin; the methodological difficulties and potential threats to validity and reliability are real; and the community of interest at this time is relatively small. Despite these challenges, we feel that increasing our understanding of the impact of museum experiences and of museums themselves over longer time frames would add significant value and is thus worth the effort.

Invitational Gatherings

We believe the field would benefit at this juncture by drawing together researchers and practitioners in order to collaboratively resolve many of the key issues surrounding long-term investigations of museum visitor impact. Following are some of the definitions, range of interests, time frames, influences, methodological challenges, and theoretical frameworks that such an invitational gathering might work on.

Broader Definitions

A broad, multidimensional definition of impact may help counter the inherent tradeoffs in setting goals and attempting to assess the impact of museum experiences. Whether consciously acknowledged or not, exhibitors, programmers, and researchers roughly make choices between (1) documenting clearly defined outcomes immediately or soon after the visitors' experiences, and, on the other hand, (2) documenting sustained or persistent, and sometimes evolving, open or nondefined outcomes in several weeks or months after an experience, or even much later in the visitors' lives. Each of these contrasting approaches is suited to different needs. For example, near- or short-term assessment make utilitarian sense for both down-and-dirty formative and more carefully constructed summative evaluations. On the other hand, intermediate- or long-term assessments are better suited to questions of how memorable the experience and how profound the learning was, and how it was embedded into the visitor's biography. How we address these two approaches (or even mix them in our evaluation and research studies) has implications for our choices of learning and research goals and strategies, and even for the scope of the underlying ambitions of our programs and organizations.

Shared definitions enable better understanding within a community of practice. In order to communicate and agree on goals and research strategies, we need to address definitional questions such as: What are the functional meanings of "long-term impact" (categories) versus "long-term outcomes" (individuals)? Are "long-term impact," "long-term memories," "long-term learning," and "long-term meaning-making" functionally equivalent, and if not, how are they different? Would it be useful to adopt an expanded three-level nomenclature of "immediate assessments" (during visit), "intermediate-term assessments" (weeks or months later), and "long-term assessments" (years later)?

Larger Units of Analysis

Is there utility in expanding the unit of analysis beyond individuals, to include families, school classes, and communities? Assessing impact on these larger units of analysis, collected at some distance from the museum experience, is complicated, but can add important information and alleviate some of the data bias that occurs through selective remembrance. Based as they are primarily on discrete memories of individuals, how can these individualized and dispersed recollections be aggregated for meaningful analysis (other than, of course, by interviewing groups)? Some interesting conclusions have been drawn from reports of widely shared iconic memories (e.g., visits to world fairs), but other research strategies may have to be developed to make sense of the long-term impact of the plain-vanilla group museum visit. For example, does the typical school group visit share enough common and memorable elements that interesting questions of long-range impact can be studied? Do the recollections of family vacation trips to historic sites and national parks hold promise for useful analysis? Do visits to museums plant enough common, long-term memories that can be analyzed to get a handle on the broader impact of museums in the community beyond evidence of individual outcomes?

Longer Time Frames

Studies that utilized intermediate-term assessments (time frames of several weeks or months), for practical reasons, make up the majority of the slender body of long-term impact research. Most of the information that can be recovered weeks or months after the event is different from the memories of experiences that are so vivid and transformative that they can be recalled in substantial detail years after the event. Longer time frames are likely to act as powerful filters, leaving in memory the most important aspects of an experience in ways that add value to the field by uncovering significant positive or negative remembered experiences that may yield other and possibly profound evidences of the impact of museums.

Methodological Clarity

If we aspire to study and make sense of memories as a way of confirming the impact of museum experiences over longer periods of time, we will need to develop and become comfortable with a wider arsenal of longitudinal research tools. Particularly intriguing are the possibilities of comparing years-later, open-ended interviews, recorded narratives, and written essays with verifiable archival material (photographs, catalogs, drawings, planning documents, press coverage) that might establish how closely an individual's memories paralleled the objective evidence, and particularly how the substance of what was learned was reflected in the goals of the original exhibit or program developer. Future studies ought to provide stronger rationale for the choice of research design and methods, and ought to link to the breadth of existing studies to allow for better compatibility of findings.

Influence of Memory and Subsequent Reinforcing Experiences on Immediate Experiences

Distant memories and their influence on immediate experience, and the role of subsequent reinforcing experiences on affective and cognitive learning, need to be better integrated in biographical approaches to museum assessment. There are intriguing reports that memories of earlier museum experiences, sometimes long buried, when triggered by later events, jell or are reorganized and given new or deeper meanings in the recovery process. These "aha's!" (when latent knowledge is finally embedded into a conceptual framework through one additional learning experience) mark new learning based on very strong but not fully realized impressions, and are often the most powerful "learning experiences" of visitors. They are also an important aspect of free-choice learning. Among the avenues for assessing long-term impacts of museum experiences, these may be the most productive directions for further study. Examples of these delayed insights were offered more than a few times when museum professionals were asked to tell stories of "significant, memorable, pivotal museum learning experiences" they observed or participated in (Spock, 2000a, 2000b). Museum visits are not isolated events, and researchers ought to distinguish between three very different effects of museum visits: (1) museum experiences can reinforce prior experiences, (2) they can provide new experiences, and (3) they may spark new experiences. All three aspects are intertwined as individuals meander through their own biographies.

Theoretical Frameworks

Many early studies that examined learning and long-term impact have been conducted atheoretically, that is, without a grounded linkage to the theories of learning, psychology, or memories. Nowadays, there are many sound theoretical platforms backed by hundreds of studies upon which the outcomes of long-term impact studies can be interpreted. For example, constructivist learning theories (of which there are various kinds) permit interpretation of learning both in informal and formal contexts (see, e.g., Anderson et al., 2003); schema theories of memory can be employed to understand how visitors recall events and how they "encode" memories; and recent research about links between affect and long-term memory can help in determining factors that manifest memories in visitors (Anderson & Shimizu, 2007). Finally, the Integrated Experience Model (Storksdieck, 2006) and the Contextual Model of Learning (Falk & Dierking, 2000) provide frameworks that specifically embed museum visits into previous experiences and follow-up and subsequent reinforcing experiences.

CONCLUSIONS AND IMPLICATIONS

The investigation of the long-term impact of museum experiences is a wide, largely underresearched field in which there is ample opportunity and scope for new investigations to contribute significantly to the understanding of visitor experience. Long-term impact studies may not only provide the field with a more complete understanding of the benefits visitors derive from museum experiences, they may also help the museum field better understand the true value of museums for the communities they serve. What is currently needed is a broad, comprehensive longitudinal impact study that follows visitors of diverse museum types (science, art, history, identity, etc.) over a long period of time, using a variety of measures, to document impact; a study akin to similar ones conducted in the medical field. Thus, deepening our understanding of the long-term meaning of museum experiences could have huge implications for the museum and leisure study field in general. This understanding could help museum professionals in developing institutional missions and goals, designing exhibitions and programming, and assessing visitor experience and learning.

There is a range of implications for museum practice, evaluation, and research, some of which we briefly summarize as a starting point for future conversations.

A. Implications for Practice

- *Align missions and goals with biographical perspectives of visitors*
  Museum visits are part of a large free-choice and leisure infrastructure, and visitors may not perceive them as unique and individual events in their lives. Museums ought to acknowledge their role within this broad tapestry of experiences.
- *Design with past and future experiences in mind*
  Linking to previous experiences and providing opportunities for follow-up are already part of good museum practice.
- *Create partnerships to link museum visit to other experiences*
  Linkages are best achieved when cultural institutions cooperate. Competition is healthy, but museums ought to reach out to other institutions that operate within their "educational" or "experience" infrastructure to enrich their visitors' lives and to provide more opportunities for subsequent reinforcing experiences and follow-up.
- *Encourage repeat visits*
  Part of any subsequent reinforcing experiences and follow-up are return visits. These may be more likely if visitors are being provided with individualized experiences that fit into their biographies.
- *Provide opportunities to remember*
  Visitors will remember some aspect of their museum experience. Museums ought to strategize what it is that they want their visitors to most remember and then encourage visitors to experience these things. Naturally, key memorable experiences will differ for different visitor types, and will range from object memories to social memories.
- *Be aware that visitors remember the visit selectively*
  One strong incentive for "managing" visitor memories is the fact that visitors might otherwise recall the unexpected: something negative that stood out from the expected. Hence, museums need to avoid negative

associations at all cost. They will otherwise color visitors' perspectives, or—as suggested above—balance them with easily obtainable positive memories.

B. Implications for Evaluation

- *Include long-term in logic model*
  This should happen as a matter of principle. Only if long-term goals are part of the design phase for visitor experiences will they become default aspects of evaluation.
- *Define outcomes over time*
  Currently, outcomes tend to be defined as if they occurred or become evident at one moment. It would be helpful to distribute outcomes along a timeline that acknowledges more openly the difference between immediate, intermediate, and long-term outcomes of museum visits.
- *Ensure realistic timelines for including longitudinal components in evaluation and link those to a theoretical framework*
  Most projects do not allow for longitudinal research because evaluation studies need to be completed shortly after programs or exhibitions are made available to the target groups. The funding community ought to decouple project funding from evaluation funding to allow for independent timelines or extend the timeline.
- *Use multiple methods and ensure that methods don't create a bias*
  No matter what the time frame of an evaluation, laser-like studies on predefined outcomes will always benefit from additional open-ended questions, and open-ended research will often benefit from using comparable indicators.

C. Implications for Research

- More research is needed to link memory and impact, and we need a refined model of learning and memory over time that is applicable to free-choice settings.
- We need to assign meaning to certain time frames and create consensus in the research community on what these time frames represent. Does it matter whether we ask visitors to talk about their museum ex-

perience three months, a year, or 30 years after a visit? What questions are best asked within what time frame? Is there an "optimal" time delay for sampling "intermediate" and "long-term" experiences?

- We need to include more biographical research to make stronger links between prior experiences, museum visits, and subsequent experiences.
- We need to expand the research methods currently in use to include ethnographic studies of communities around museums, or conduct more quantitative, multi-institutional studies of long-term learning over specified time frames.
- We need to better understand the links between immediate, intermediate, and long-term impacts.

ENDNOTES

1. See Ballantyne & Packer, 2005; Borun et al., 1996; Dierking, Luke, Foat, & Adelman, 2000; Doering & Pekarik, 1996; Falk, 2006; Hood, 1983; Moussouri, 1997; Paris & Mercer, 2002; Pekarik, Doering, & Karns, 1999; Prentice, Davies, & Beeho, 1997; Rounds, 2004.

15

Investigating Socially Mediated Learning

Tamsin Astor-Jack, Kimberlee L. Kiehl Whaley, Lynn D. Dierking, Deborah L. Perry, and Cecilia Garibay

Learning is the process of acquiring knowledge through experience. This can occur in a variety of settings (e.g., school, museums) and within a variety of sociocultural contexts (e.g., alone, one-on-one, group). Since learning is so complex, researchers generally focus on specific aspects of learning, for example, where in the brain learning occurs.

In this chapter we focus on the *processes* of learning—specifically, socially mediated processes—rather than discussing the outcomes of learning in museums. We feel that this focus is warranted since most people visit museums in social groups composed of people with varied interests, temperaments, capacities, and backgrounds (Pott, 1963). During their visits, groups engage in social interactions taking place (1) on various levels (adult-child, peer-peer) and (2) across many groups (among visitors, or visitors and staff).

We argue that understanding the role of the social processes of learning is essential to understanding the nature of learning in museums. Taking a sociocultural approach provides a broad frame for considering how the visiting group is situated within a wider cultural/social context and also highlights learning processes at the group or community level—beyond the individual learner (Ellenbogen, Luke, & Dierking, 2004).

When reviewing sociocultural theoretical frameworks, a challenge emerges: the study of museum learning draws on theories from many fields, such as anthropology, psychology, and education. Thus, a one-to-one mapping of theory to research, and research findings to practice, is often not directly applicable. To address this conundrum, we focus on two sociocultural theories and attempt to provide links between different theories and this research and their implications for practice. We then discuss the theories within the context of *what we know.* In the final section we highlight areas of research that we feel would help professionals in the field to understand more effectively social mediation as a tool for supporting learning in and beyond the museum.

THEORETICAL FRAMEWORKS

We have selected two central ideas as a focus: (1) mediation and Vygotsky's zone of proximal development and (2) Rogoff and Lave's concept of the community of learners.

Mediation and the Zone of Proximal Development

Vygotsky considered development to be an interaction between one's current and emerging abilities *and* one's social situation. He proposed the zone of proximal development (ZPD): the difference between one's "actual developmental level as determined by independent problem solving" and "potential [level of] development as determined through problem solving under adult guidance or in collaboration with more capable peers" (Vygotsky, 1978, p. 86). A key feature of the process by which a learner moves from where he is to where he is capable of being is social mediation, or scaffolding (Wood, Bruner, & Ross, 1976). During scaffolding, a more experienced "tutor" provides guidance and direction that help the learner move through her ZPD. Vygotsky considered the convergence of speech and practical activity (Vygotsky, 1978) and speech and thought (Vygotsky, 1986) to be vital in cognitive development. These ideas are particularly pertinent to understanding the socially mediated processes of learning in museums, since the museum experience *is* the practical activity accompanying the speech that occurs between visitors as they interact and think about the experience. In many situations, this social mediation encourages the individual to broaden her ZPD and develop beyond her capacities; and serves as a form of distributed meaning-making since understanding often resides within the group, rather than any one individual.

These concepts are not limited to children's interactions with adults or more advanced peers—it can, and has been, applied to adults who are in new learning environments. For example, Rudman, Sharples, and Baber (2002) examined adult conversations and found that by providing a computing device that scaffolded their current understanding, individuals learned more, compared to those with pen and paper or a computing device that did not scaffold their understanding. Additionally, children can scaffold adult learning if they are the more capable partner—often the case in exhibitions with advanced technologies.

Community of Learners

Rogoff and Lave (1984) suggest that all knowledge, including the specific knowledge of any group or society, is constructed within a sociocultural context; that is, knowledge is not the same for all individuals in society but is shared within delimited communities of knowers. They suggest that all learning occurs within a context they call a "community of learners" and that a myriad number of communities of learners exists, defined by the boundaries of shared knowledge and experience, for example, the family or a special interest group (e.g., an astrology club). Within each group a set of shared beliefs, values, language, and practices exists, what Gonzalez and Moll (2002) call "funds of knowledge." Learning occurs, therefore, when "new generations collaborate with older generations in varying forms of interpersonal engagement and institutional practice" (Rogoff, 1998); thus, social mediation is one of the main processes by which we learn.

Membership in a community of learners can either be conscious, when a child aspires to become an astronaut and does everything possible to achieve that goal, or unconscious, when an individual "independently" pursues an interest or avocation only to discover that there are many others who also share a similar interest.

WHAT WE KNOW

Watch museum visitors for any length of time and one thing becomes very clear: people do not simply visit exhibitions, they talk with each other as they explore. Moreover, these conversations are important learning tools for higher levels of understanding. Baxandall (1987) says it well: the art museum experience was not about "looking at pictures but about talking about looking at pictures."

But what's going on as visitors are talking about what they are looking at and experiencing? It is encouraging to note that more and more research focused on the role of social mediation and the development of communities of learners in museums is being conducted.

The Role of Mediation and the Zone of Proximal Development

A common type of interaction that is associated with educational experiences is *talking at* others, commonly observed in formal educational settings, but also a characteristic of what goes on in museum settings. Here, a more knowledgeable person takes an active role in organizing and "delivering" the learning experience, while the others are passive recipients. In museum settings parents or teachers might read labels or tell the child what to do or how the particular exhibit component works. While an effective teaching/learning strategy in many situations, both formal and free-choice, *talking at* and *explaining* is only one of many strategies museum visitors use as they engage in socially mediated learning.

Many studies highlight the importance of social interaction among family members in museums, particularly between adults and children, where parents are frequently found to interact in directed ways with their children. In many learning experiences, there tends to be a clear mentor-novice relationship with parents often assuming the role of "teacher" (Diamond, 1986). For example, 81% of children's time in the *Launch Pad* at the Science Museum of London was spent interacting with family members and other visitors, with the adults playing the roles of interpreter, teacher, and explainer (Stevenson, 1991). An interesting aside is that these interactions among family members are also affected by gender. Crowley, Callanan, Tenenbaum, and Allen (2001) discovered that if the exhibition required no explanation, parents talked to boys and girls with the same frequency and in the same way. However, at exhibitions requiring explanation, parents were three times more likely to offer a scientific explanation to boys than to girls, regardless of the child's behavior.

In addition to the strategies of explaining or talking at, in many situations group members work together to jointly construct understandings—*talking with* each other. These interactions tend to be somewhat balanced, with the mentor role shifting among the people engaged; in some cases there is, in fact, no clear mentor. The majority of research in this area involves the analysis of conversations among visitors in a group, focusing on the roles members assume, the strategies used, and which topics are discussed.

Most of the research on socially mediated learning in museums focuses on parent-child interactions. Parents can be effective facilitators for their children's learning if exhibitions are designed with collaborative learning in mind (Borun, Chambers, & Cleghorn, 1996; Borun, Chambers, Dritsas, & Johnson, 1997; Dierking, 1989; Dierking & Falk, 1994; Perry, 1989). However, adults must feel comfortable with the content and experiences in the museum. Parents' perceptions of a discipline influence how they facilitate their child's learning and convey process and the epistemology of knowledge (Luke, Coles, & Falk, 1998). In another study, parents actively used inquiry skills, particularly observing and questioning, when interacting with their children (Ash, 2003). Additional research shows that parents willingly engage in active learning (in-house library, family activity kits, etc.), if they know about and understand the role of materials in providing assistance to their children (Falk & Dierking, 2000). Adult-child interactions also occur in school group visits as teachers and chaperones interact with children (e.g., Sedzielarz, 2003).

Peer interactions are less well documented. They tend to be more equitable, with a less well-defined mentor-novice relationship. Children who experienced an interactive science exhibit in a group were twice as likely to use the exhibition as intended than children experiencing it alone (Rennie & McClafferty, 2002). Further, Gottfried (1979) found that the most meaningful exploration occurred in groups of two.

Having stated the importance of the social dimensions of learning, as a field we are still trying to understand the complexities of its role, since the types of behaviors observed at exhibitions (e.g., reading, manipulating, and talking) vary according to the composition of the social group (see, e.g., McManus 1987, 1988). What is considered optimal meaningful social interaction within an exhibition or a program will vary depending on the composition of the group (e.g., culture, gender, size, and age), whether visitors are seasoned or novice visitors, and so on. For example, U.S. Mexican-heritage children simultaneously attend to multiple events more than European-American-heritage children who alternate their attention between events (Correa-Chávez, Rogoff, & Mejia, 2005). Also, Garibay, Gilmartin, and Schaefer (2002) found that novice visitors initially needed more staff facilitation regarding exhibitions. Clearly this is a complex issue.

Unfortunately, there is a dearth of studies that have systematically examined potential cultural differences in social interactions in museums. Some groups prefer to visit in larger or extended family groups (Falk, 1993; Garibay et al., 2002), which may create different dynamics that have not been fully explored. For example, Garibay (2004) noted that Latino families who came to an exhibition together engaged in "cross-parenting" where adults took on and traded the caregiver roles for their friends' children.

Socioeconomic factors may also influence socially mediated learning, particularly around the types of teaching roles adults may take. For example, Rogoff, Paradise, Mejia Arauz, Correa-Chávez, and Andelillo (2003) suggest that parents with higher levels of educational attainment may determine the degree to which they use "school-like discourse formats," those which are based on transmission of information from experts. Currently, most museum visitors tend to be well-educated; suggesting that what we know may be limited by those we are studying—again pointing to the importance of being self-reflective about our assumptions of the role of the socially mediated processes of learning.

Community of Learners

In thinking about how visitors learn in socially mediated ways, it is also helpful to examine the ways in which they form their own communities of learners, groups that, through social interaction come to common understandings of the topic. Further, groups such as families or peer groups are by definition already a community of learners that have selected the museum as a place in which to interact. While the research in this area in museums is sparse, classroom research has demonstrated that children engaged in peer interaction form understandings via the creation of communities of learning.

Iwasyk (2000) examined the use of questions and communication skills during her elementary students' conversations about science. She determined that children needed to ask questions, that asking and answering questions helped clarify their thinking, and through this "teaching" the students learned. Gallas (1995) moved her classroom from questions based on the teacher agenda to ones led by the children; and discovered that only when she stepped away and let the children lead the questioning could she see children thinking together—co-constructing ideas and forming theories. Matusov, Bell, and Rogoff (2002) also report that during joint problem-solving, children built a community and worked collaboratively. These studies highlight the importance of flexibility within the learning environment such that shared meaning-making is highlighted.

There are few studies that have taken this approach to investigate museum learning specifically. Gilbert and Priest (2001) observed peer-peer interactions among 30 fourth-graders at the *Food for Thought* gallery in the Science Museum of London. Students formed their own communities of learners as they talked, and these interactions played a major role in group learning. Further, while critical incidents (e.g., a surprise action) initiated discourse, the continuation of discourse and the joint meaning-making was facilitated by links made between the museum experience and the outside lives of students.

Such findings have made researchers appreciate that it is not enough to study simply the learning interactions within the walls of the museum, particularly as these groups are already a community of learners, inside the institution and outside. Research is being extended to include learning beyond the museum walls, such as follow-up interviews with families weeks, months, or years after the visit. Other researchers are exploring think-aloud techniques, journaling, and in-home interviews (Ellenbogen, 2002; Falk & Storksdieck, 2005). Researchers have repeatedly shown that many of the conversations that begin in the museum continue in the car or back home (Ellenbogen, 2003; Falk, Luke, & Abrams, 1996; McManus, 1993; Stevenson, 1991). In other studies families described specific exhibit and program elements without prompting, indicating the durability of the museum experience, and discussed connections between the content in the exhibition and other circumstances or phenomenon, demonstrating the integrations of these experiences into their daily lives (Luke et al., 1998, Luke, Büchner et al. 1999).

FUTURE DIRECTIONS

In this chapter we have argued that two social dimensions of learning are critical in conceptualizing and designing research and practice in the field and ultimately should also shape policy decisions. Few in the field would dispute the rhetoric that learning is socially mediated and that most people visit museums in social groups. We are still unclear as to what constitutes optimal social interaction within an exhibition or program, for whom, and under what conditions, nor do we fully understand nonverbal interaction, emotion, and other social processes in this context. This is vital as more underrepresented groups start frequenting museums so that we can research and then accomplish their free-choice learning needs. What do we need to do to move from the rhetoric of social mediation to more informed research, practice, and policy in the future?

Research

Up to now, we have made progress in identifying appropriate theoretical frameworks within which to ground our research focusing on sociocultural theories, guided by the early work of Piaget, Dewey, and Vygotsky and furthered through Rogoff and Lave. However, there are still areas in which further exploration is needed to maximize the social aspects of learning in museums (Falk & Dierking, 2002), for example, research that could help practitioners understand effective strategies for facilitating adult-child and peer-to-peer interactions among visitors of all ages and types of groups (e.g., families, school groups, tourists, and teenagers on a date).

As we have argued, it is critical that research in this area includes diverse audiences. More research needs to be done on the role of nonverbal interactions, particularly among groups with young children and those with English as a second language. Cultural differences extend beyond the issue of language (Correa-Chávez et al., 2005; Garibay, 2004).

Ultimately, we need to understand the learning experience from the visitor's perspective, rather than the researcher's. We need to provide an effective forum for hearing visitors' voices, including children, who are an important museum audience (Children's Voices, 2005), an approach that has been attempted in formal education (Elmesky & Tobin, 2005). Studies focused on after-school and youth-based programs are sorely needed in museums.[1]

Given the importance of conversational techniques and a focus on inquiry, the field could benefit from research on effective types of questioning and scaffolding. Funding for applied research could be invaluable, for example, investigating which techniques work best for which types of social groups and under what conditions.

There is also a need for precise definitions/terminology, for example, clarifying similarities and differences between *family learning, intergenerational learning*, and *peer-peer learning.* Research is limited by methods and tools, usually designed for individuals rather than groups. We need to use the group, not the individual, as the unit of analysis. Ultimately, we need to investigate how a group is situated within wider cultural and social contexts, highlighting learning at the level of the visiting group and community. Research focused on the audience and its active construction of meaning is a creative process consisting of interactions among the audience, individual objects and exhibitions, and the institution as a whole (Silverman, 1990). Making progress in any of these areas would benefit the field in the future.

With a few exceptions (Sykes, 1992, 1993), the influence of more open-ended, less directed exhibitions on family interactions has not been investigated and little attention has been paid in the context of the social dimensions of learning to those adult-child interactions in which the child is the more skilled partner.

Institutional Practice

Progress has also been made in the area of practice, particularly in exhibition and program development where there has been some effort to embed socially mediated notions of learning into the design process. In other words, we are getting better at *talking with*, rather than *talking at* visitors and more effectively facilitating visitors' engagement with exhibitions and programs to build and shape their identities and interests (Perry & Morrissey, 2005). However, more changes are needed and it is important for us to document both successes and failures in this area. We must also advocate socially mediated approaches as essential to exhibition and program design, not just as "fun" touches. Though practitioners appreciate that social interaction is vital to personalizing learning, most interactions between museum staff and the public remain didactic. Interactions between members of social groups are often "tolerated" rather than explicitly designed for (Falk & Dierking, 2002).

We also need to think beyond the physical space of exhibitions and programs. The future will include customization and personalization (see chapter 6 in this book). One avenue already being explored is in the arena of technology, for example, cell phones and personal information devices. Museums need to be key players in this arena and strive to ensure that as these technologies are developed as devices to further learning, they are designed to facilitate social interactions and mediation, as well as independent learning. In addition, museums have an opportunity to position themselves as important community resources for diverse learners and in doing so the institutions become learners themselves. Museums are beginning to understand the communities' needs, interests, and identities, and hopefully this will continue to be a fruitful area for practice as museums learn to capitalize on their visitors.

There is also a need for stronger connections between practice and research. Progress has been made thanks to varied efforts and funding focused on encouraging and formalizing such interactions between research and practice institutions.[2] Researchers interested in influencing practice need to learn to communicate in succinct, clear, and less jargon-laden ways. Unfortunately, despite a large investment in the past ten years, many of our research efforts have not yet generated both compelling *and* usable models and ideas. Communicating what we are learning to both practitioners and researchers is critical to future improvements in practice.

Policy

Better understanding the social nature of learning also has important policy implications for both the field and society. Exhibitions and programs could also serve as forums and conveners for the discussion of tough societal issues that the neutral ground of a museum can accommodate.

The field also needs to understand the barriers to social learning in museums, particularly for people who have not historically used museums as resources. Issues to consider here include making the museum experience more amenable to underrepresented groups or providing museum staff to support novice visitors. We need to explore models for listening to communities directly and reevaluate the "outcome" measures used as evidence for learning, to ensure that they reflect the social/contextual nature of museum learning.

However, the ability to understand better and articulate the social nature of learning in and from museums is also critical to general educational policy in the future. Worldwide, societies are at an important turning point—a "tipping point" (Gladwell, 2000)—the dramatic moment when an idea, trend, or social behavior crosses a threshold and then spreads. This trend we are observing is a transformation in understanding how, why, where, and with whom people learn. Worldwide there is increasing dissatisfaction with the monolithic institution—the school—and a beginning appreciation for the idea that all parts of society have an important role to play in the educational enterprise.

This is the moment for museums to articulate and provide evidence for their role in facilitating lifelong, free-choice, socially mediated learning. Museums are resources for children visiting from their schools, and they also serve the learning needs of teens, families, and seniors. The future requires that we have a better educated citizenry, beyond merely being able to pass multiple-choice tests or demonstrate skills necessary to be a productive and successful person in the workplace. For example, if one becomes a parent, it is hoped that he will have a better understanding of how to facilitate the learning of his children and others in his immediate family. The citizen will also hopefully benefit from an awareness of the rich worlds of experience and knowledge that represent the collective million-plus years of human history, as well as the skills and capacity to explore her own personal, intellectual, and spiritual growth and enrichment—in other words, to continually learn wherever he is, whenever she chooses, with whomever, throughout a lifetime.

As a field we know that museums are excellent places in which to fulfill these learning goals, representing key resources for lifelong, free-choice learning that can be powerful and memorable. We know that there is evidence that such learning can support lifelong hobbies, encourage career decisions, and teach people the joy of learning, yet most citizens (and many of our elected policymakers) are not so aware. Thus, we need to continually strive to be essential to the life of our communities and develop better ways of sharing and communicating our activities and research findings within our own field, but also ultimately with policymakers and citizens so that policy decisions reflect support for our critically important institutions. It is by better understanding, facilitating, and building upon the social learning interactions in museums that we will be able to make our case more strongly.

ENDNOTES

1. One such effort is underway with NSF funding. McCreedy and Dierking are utilizing a community of practice approach to conduct a retrospective study of the long-term impact of in-depth gender-focused programs in science, tracking down young women, many from underrepresented communities, who participated in programs over five years ago. Findings hopefully will yield useful information about the role that a community of practice can play in supporting girls' science learning.
2. See the Institute of Museum and Library Services–funded Museum Learning Collaborative, National Science Foundation–funded Centers for Learning and Teaching, informalscience.org, organizations such as the Visitor Studies Association, American Association of Museums and Association of Science-Technology Centers, and so on.

16

Research in Museums: Coping With Complexity

Sue Allen, Joshua Gutwill, Deborah L. Perry, Cecilia Garibay, Kirsten M. Ellenbogen, Joe E. Heimlich, Christine A. Reich, and Christine Klein

Museums and other free-choice learning environments are challenging to study. One reason is the enormous variability in both the environment and the audience—researchers need to address a multitude of interacting factors in order to make sense of what visitors learn. Falk and Dierking's Contextual Model of Learning (2000) underscores the complexity inherent in visitor studies by identifying four broad contexts or influences, each of which is a complicated world unto itself: the physical, personal, and sociocultural realms, and time. Their complexity and interrelatedness can easily overwhelm a researcher. For example, how does one characterize a "typical visitor experience" even in terms of basic exposure, if each person's attention takes a unique path through a densely featured space of possibilities?

Fortunately, museum researchers and evaluators have adapted and developed methods for managing the complexity involved in studying free-choice learning. In this chapter, we categorize such methods as falling into two main approaches: reducing complexity to identify causal relationships, and embracing complexity to create a deep understanding from multiple perspectives. These two approaches have deep roots in social science, and both are practiced by well-respected researchers (National Research Council, 2002). For each, we mention some of the underlying philosophical assumptions, show how the approach applies to visitor studies in particular, and highlight (with examples) some useful methods for maximizing quality. We do not advocate one approach over another;

rather, we attempt to highlight some of the issues that arise in the challenging processes of study design and implementation.

REDUCING COMPLEXITY TO IDENTIFY CAUSES AND EFFECTS

Museum practitioners design learning environments intentionally, with some form of goal or purpose (however broadly defined). As a result, they often desire an approach that searches for cause-and-effect relationships to determine how well their goals are met and how their designs affect visitors. This is also desirable to those researchers and evaluators who believe in the possibility of general principles that characterize which environments and experiences will result in which outcomes.

Cause-and-effect research such as randomized clinical trials is treated in some circles as the "gold standard" of social science research because of its predictive and explanatory power (see Shavelson & Towne, 2004). However, it is not often used in its pure form in visitor studies for two main reasons, one theoretical and the other logistical. The theoretical question is whether it is possible to identify and measure key concepts such as motivation, engagement, or learning, and to assume they mean the same thing for a variety of visitors in a variety of situations. In a field where so much emphasis is given to visitors' personal interpretations and choices, can one legitimately combine different people's experiences and draw conclusions about characteristics of visitors and their learning? Researchers who support experimentation believe that this is indeed possible, and they point to social psychology, which has a history of taking complex situations in real-world settings, pulling out a small number of variables for study, and arriving at fundamental principles of social behavior that can then be applied back to the real-world situation. For example, a baffling case where a woman was murdered in full view of the windows of a large apartment building led to a set of laboratory experiments by Latane and Darley (1969) on altruistic behavior, ultimately yielding an explanation of why no one had called for help at the time.

Such experiments need to be done carefully, however, to have any credibility, and certain controls and comparisons are needed to rule out competing explanations. This leads to the second reason that the field of visitor studies lacks controlled experiments: the logistical challenges. Often researchers do not have the ability to "randomly assign" visitors to different versions of an exhibit or program, yet random assignment is a fundamental principle of this kind of work. Similarly, the experimental approach relies on comparisons: between visitors who were exposed to a program and those who were not, between visitors before and after using an exhibit, and so forth. Such comparisons are often expensive and difficult to arrange, and may intrude on visitors' natural behavior.

In spite of these challenges, some experimental research is being done in museums, and the principles can even be incorporated into regular evaluation procedures when the goal is to make causal claims about learning. Here we describe three ways to reduce complexity and increase the likelihood of identifying cause-and-effect relationships in free-choice learning studies: (1) simplify effects into a few measurable outcomes, (2) focus on a small number of potential causes, and most importantly, (3) reduce the number of competing explanations.

Focus on a Few Measurable Outcomes

Outcomes (or "dependent variables") are measurable aspects of visitors that may have been influenced by their experiences. They might include such things as knowledge, skills, behaviors, attendance, emotions, and so on. One way to simplify an array of possible outcomes is to select certain aspects of the output and ignore the rest. For example, rather than analyze entire visitor conversations, Humphrey and Gutwill (2005) chose to study only the questions visitors asked themselves and the statements that immediately followed, because one of their project goals was to shift authority from the exhibit's label text to the visitors' own driving questions. Similar selection of key outputs happens every time researchers force visitors' responses to fit into a predetermined numerical scale, and even interviews are best analyzed with respect to a dimension or variable that is motivated by theory or stakeholders' questions.

A second way to simplify outputs is to give something a single score that captures the "big picture." For instance, rather than identify a dozen different aspects of visitor inquiry at a science museum, one could assign a single inquiry score to each visitor group, based on overall judgments of their behavior. Properly done, this kind of holistic scoring is easier to analyze and has been found to be just as effective at characterizing visitors' inquiry behaviors (Randol, 2005).

Finally, there are statistical techniques for simplifying the outputs after a study has been run, such as cluster analysis and skree plotting. Cluster analysis is a process for finding patterns by clustering respondents into meaningful groups (e.g., Hui-Min & Cooper, 2001). Skree plotting and other scatterplot methods provide a visual and statistical means for observing the logical groupings of variables (e.g., Heimlich, Storksdieck, Barlage, & Falk, 2005).

Focus on a Small Number of Potential Causes

When trying to understand the reasons behind an outcome, it is often helpful to limit the number of potential causes ("independent variables") included in the study. Having fewer variables improves the statistical chances of finding a result if one exists, especially important in visitor studies that involve small samples of visitors. It also tends to improve the clarity and interpretability of the data analysis (Prosavac, 1998).

Factors that might affect the outcomes but are not the focus of a study can be dealt with in three ways: by fixing them at a single level, by randomizing, or by blocking. For example, if age is expected to affect visitors' experiences, but this is not the focus of the study, a researcher could fix this variable by including only visitors whose ages fall within a narrow range. Alternatively, visitors of all ages could be assigned randomly to different versions of the offering being tested, so that after many visitors have participated, the effect of age would "wash out," being roughly equal across all versions. Or, using the third technique, a researcher could "block" the variable into categories such as child, adolescent, and adult (ideally, natural groupings observed in pilot studies). Visitors would participate until a representative number of visitors existed in each block. These three techniques occupy different positions in the trade-off between deeper understanding of a particular group's experience on the one hand, and broader applicability of the study results on the other. Finally, factors that cannot logistically be controlled for using these techniques can often be dealt with statistically following data collection.

While reducing the number of variables in a study is an effective way to cope with complexity, such simplification comes with two cautionary notes. First, it is always possible that the learning outcomes observed may actually be due to a variable that was not measured or recognized as important. Second, the approach puts limitations on the degree to which the results can be applied to other situations. For example, researchers studying exhibit narratives at the Exploratorium were so concerned about limiting the number of variables that they ended up studying narratives that lacked aesthetic variation and interest; later, they revised their study design to allow for testing of narratives that were more representative of realistic museum creations (Allen, 2004). Even something as simple as fixing the exposure to an exhibit label by asking visitors to read all of it may change the context of learning from that on the open museum floor. Researchers frequently face this kind of trade-off between the rigorous design ("internal validity") of an experiment, and the degree to which its results can be generalized beyond those specific circumstances ("external validity"). Often they compromise, limiting the number of variables but choosing them carefully to capture a few key causal relationships and not drift too far from the real-world context of museum learning.

Reduce the Number of Competing Explanations: Some Useful Study Designs

A key characteristic of an experimental study is that it allows researchers to rule out alternative theories or explanations of an observed outcome. Only rarely are visitor studies created with this as a priority. Usually, limited evaluation resources are focused on characterizing outcomes of interest (e.g., visitors' behaviors, or their reflections on a visit or program) rather than proving that these were generated by specific causes (e.g., proving that visitors' understandings came from experiencing a program rather than from their prior knowledge, or that visitors' increasing engagement at an exhibit was due to improvements in its design, rather than changes in the population of visitors using it). However, in an era when museums are increasingly being asked for proof of their impact and the factors that contribute to it, we list several examples of ways to adjust some of the most frequently used research and evaluation designs to support more rigorous cause-and-effect inferences. We draw on well-known overviews given by Campbell and Stanley (1963) and Cook and Campbell (1979).

Interviewing After an Experience

A common evaluation design is the exit interview, in which visitors are interviewed as they leave an exhibit or program, and conclusions are drawn about what they learned. While this design can efficiently capture visitors' reflections about what they did and thought during their experience, it usually relies on memory and verbal self-report of what has been learned, rather than directly assessing any change in visitors' knowledge, attitudes, or skills. Unfortunately, an exit interview that directly assesses such things as visitors' knowledge as they leave the program exhibition is vulnerable to the competing explanation that the visitors already had such knowledge and abilities before they arrived.

Interviewing Before and After ("Pre Versus Post")

To reduce competing explanations, a somewhat stronger version of the exit interview involves interviewing visitors twice. For example, in personal meaning-mapping (Falk, Moussouri, & Coulson, 1998), visitors represent their knowledge on concept maps before and after their visit. However, the design is still vulnerable to alternative explanations: learning might have been enhanced because the visitors knew they would be interviewed (e.g., Serrell, 2000), or because they got practice doing the assessment. Also, it is difficult to design pre-interviews that fully anticipate the kinds of learning an individual visitor might do, and because it takes twice as long as a single interview, it may be difficult to recruit a representative sample of visitors.

Interviewing Two Groups

Another slight improvement over the exit interview is the two-group interview. The researcher conducts only one interview, but with two sets of visitors: those who chose to explore the offering (e.g., a program) being assessed, and those who did not. Comparing the two groups gives some idea of the effects of the program as compared with the rest of the institution. However, this design is open to the competing explanation of visitor selection: perhaps the already-interested visitors went into the program and the disinterested visitors did not. This is a tricky case; museums may be legitimately reluctant to assign visitors to having or not having an experience, either because they consider it too disrespectful or because they worry that taking away visitors' choice may limit the study's applicability to the real world of the museum (e.g., how do you interpret a study that includes people who hate audio tours but learn something when forced to use one?). However, letting visitors decide means that a heavy price is paid in terms of internal validity of the study. If researchers want to prove that a particular offering resulted in a particular outcome, they need to study a large enough sample of visitors, and randomly assign them to different versions of an experience (deciding, for instance, who will get an audio tour and who won't).

Two Recommended Study Designs for High Internal Validity

Exit Interview with Random Assignment

In a rigorous version of the two-group study, visitors are randomly assigned to different versions of the experience (or one group gets nothing if the goal is to prove the experience had an impact). Only one exit interview is required because all other potential causes (such as day of the week, time of day, or a sudden busload of historians) will "wash out" across treatments if the sample is sufficiently large. This design could actually be used far more often in museum research and evaluation than it is, especially when testing minor changes in exhibits or labels where there are alternate versions that can be easily interchanged at short notice. Ideally, this could be done even in "uncued" studies (i.e., without alerting visitors ahead of time that they will be interviewed), by randomly revealing one version of the offering for discovery by the next visitor who spontaneously approaches. More realistically, if the effort of swapping versions is high, rigor may be slightly compromised by leaving each version available for a number of visitors to experience it, before changing versions. This method was used in the Finding Significance project (Allen, 2004), where cued visitors were shown one version of an exhibit, and the version changed after each family.[1]

Counterbalanced (Requires Cuing)

Another strong design is the counterbalanced design, in which all visitors see all versions of an offering, but in different orders. Gutwill (2006) used this method to test an exhibit label in three formats; visitors were asked to comment on each, and state their preference. This design rules out many competing explanations such as selection, practice effects, and so on. The main limitations are that the entire experience must be cued rather than spontaneous, the versions have to be swapped or revealed very quickly while a visitor waits, and it requires asking time of visitors to complete the whole process (leading to possible visitor selection effects).

A Note About Variability Across Interviewers and Data Coders

A final challenge to cause-and-effect experiments arises if multiple people collect or code the data. For example, one interviewer might inadvertently ask visitors more follow-up questions than another, or different coders may see different things in the same data. Even a single coder may drift over time, influencing the results in a certain direction. Most of these effects can be eliminated by having each person work equally across all variations tested, and by doing periodic inter-rater reliability tests: Multiple people record or code a subset of the data, and check their findings against one another. If the agreement drops below an acceptable level, the coding rubric is revisited (see Bakeman & Gottman, 1997). Needless to say, such testing and adjustment requires significant time.

In summary, reducing complexity is a powerful approach that can, at its best, provide evidence of the causes that lead to key visitor outcomes. Even relatively simple evaluation studies can often be strengthened by the use of appropriate control groups and comparisons, allowing evaluators to make stronger claims that learning has happened in the setting of interest, or that one design is more effective than another. At the same time, adding such comparison groups almost always uses more resources than a simple evaluation of outcomes, and the controlling of variables puts limits on the situations to which the conclusions can be applied.

EMBRACING COMPLEXITY TO UNDERSTAND MULTIPLE PERSPECTIVES MORE DEEPLY

In this section we describe methodological approaches that embrace, rather than reduce, the complexity inherent in free-choice learning settings. Naturalistic inquiry, case studies, and culturally responsive studies fall within a research paradigm for fostering deep understanding of the multiple realities of visitors' experiences, and all have been applied in museum contexts.

Naturalistic Inquiry

Naturalistic inquiry is a methodology that embraces the variability of actual environments. It is grounded in the belief that the best way to study a research question is to look at many aspects of it in as much detail as possible in the natural setting (Lincoln & Guba, 1985).

> Naturalistic Evaluation takes a broad, holistic view of the program, exhibit or institution being studied, is more interpretative than judgmental, and requires participation from a wide range of people who are to be served by the study effort. . . . The purpose is to uncover the multiple realities and multiple perspectives that exist and are provoked as people experience the museum environment—it reveals the configuration of meaning that emerges when different people are exposed to a common stimulus. (Wolf & Tymitz, 1979, p. 2–3)

Rather than search for cause-and-effect relationships by averaging across visitors, naturalistic inquiry seeks to understand the mutually influencing factors that yield a range of individual visitor experiences. This approach tends to enhance perception of the richness, complexities, and intricacies of museum environments.

Naturalistic and experimental inquiry approaches often use similar methods (e.g., interviews, observations) but the exact procedures may differ. For example, because naturalistic research is seen as fundamentally interactional, interviews often grow out of observations and previous interviews rather than following a standard protocol. Also, naturalistic inquirers tend to employ purposive sampling (selecting a particular respondent for a particular purpose), rather than sampling randomly. For example, the researcher may have just finished observing and interviewing an adult with a child, and now she wishes to see how two children working together without an adult might use the exhibit. Or perhaps the researcher overhears a particular group engaged in a heated debate at one exhibit, so asks to join their conversation. The goal of purposive sampling is to ensure that a broad range of audience diversity is included in the study, and that the interactions with any particular respondent are extended and rich, even if this results in small sample size.

This raises an important question: How does one know if a given naturalistic study is good, or at least good enough? With naturalistic inquiry, researchers employ four criteria for establishing trustworthiness: credibility, transferability, dependability, and confirmability (Gyllenhaal, 1998; Lincoln & Guba, 1985; Williams, n.d.).

Credibility

To achieve credibility, a naturalistic study should be believable to readers and approved by the people who provided the information gathered during the study. There are several procedures for enhancing credibility. During the data collection phase, *prolonged engagement* helps to build trust with respondents and allows researchers to experience variation over time. Ideally, this can be done over a period of years (Schaefer, Perry, & Gyllenhaal, 2002) but can even make a noticeable difference over a single evening (Gyllenhaal, Perry, & Cheng, 2005). *Persistent observation*, exploring details of the phenomena under study to a deep level, may reveal important or surprising results. For example, during the evaluation of a set of science exhibits at the Exploratorium, persistent observation of visitors led to the emergence of an unanticipated factor influencing the visitor experience, viz., the reasons that visitors chose to *end* their engagements with an exhibit element (Tisdal, 2004). Inquirers ought to verify their findings through *triangulation* by referring to multiple sources of information (including literature); using multiple methods of data collection, such as interviews and observations; and using multiple inquirers when possible to diversify interactions with study respondents.

After the data have been collected, naturalistic researchers engage in several procedures for checking the credibility of their conclusions. *Negative case analysis* is used to refine conclusions until they account for all known cases without exception, and *progressive subjectivity checks* require researchers to document their changing expectations for the study. During *peer debriefing*, the inquirer meets with one or more colleagues not involved in the study, who can question the methods, emerging conclusions, and biases of the inquirer, and can sometimes offer a fresh interpretation. Finally, when possible, naturalistic researchers engage in *member checking* by having their data, interpretations, and reports reviewed by the respondents to determine whether their perspectives have been adequately and credibly represented. All these procedures help to ensure that the study is credible.

Transferability

Transferability means the degree to which findings from one context or setting (i.e., where the research was conducted) can be applied to others. Whether findings can be transferred or not is an empirical question that cannot be answered by the inquirer alone; people who read naturalistic inquiry reports have to make this determination themselves. To judge the transferability of a particular study, *thick description* is necessary, meaning a clear, rich, and complete description of the time, setting, and context during which the study took place (Geertz, 1983). In practice, because of budgetary and other constraints, description is often not as thick as future readers might need, but limited to comments such as data were collected on a particularly crowded, free day.

Dependability

Dependability is the stability or consistency of the inquiry processes used over time. To check the dependability of a naturalistic inquiry, an independent auditor reviews the activities of the researcher (as recorded in an audit trail in field notes, journals, and reports) to see how well the criteria of credibility and transferability have been met. In practice, this is rarely done due to budget constraints, but dependability audits could be done by graduate students learning to conduct evaluation and research in museums.

Confirmability

Confirmability is how well the *results* are supported by events that are independent of the researcher. In studies in museums, the main method for achieving confirmability is reference to the literature or previous studies.

Case Studies

Another way of embracing complexity in museums is the use of case studies, which traditionally focus on a small number of subjects and involve highly contextualized data collection and analysis. Case studies can offset the marginalizing effects that can result simply from focusing on averages when analyzing a set of data. For example, universal design (the creation of products and environments to be usable by all people without the need for adaptation) stems from a recognition that there is no average person, and that designers should design environments that represent the diversity of size and ability that exists within the human population.

Many of the issues that are critical to conducting case studies are the same as those already described for naturalistic inquiry (in fact, case study may be regarded as a method falling under this broader umbrella). Researchers conducting case studies also face challenges that are specific to the approach, such as selecting the case and establishing its boundaries, collecting data appropriately and accurately from multiple sources over time, and interpreting and analyzing context-specific data (Anderson & Arsenault, 2004).

Selecting the Case

Arguably, the most critical challenge for a case study is the selection of the case(s). Sometimes selection is driven by a research question or problem, such as: How are museum experiences incorporated into a family's day-to-day life and ongoing learning activities? A case for such a study should not be selected to be representative but because it yields the most useful data. Specifically, a good case is cooperative, accessible, and active, and provides data relevant to the study's purpose. For example, in order to select good cases for an ethnographic case study of the role of museums in family life, Ellenbogen (2002) selected only families who went to museums six or more times per year, even though this was twice the typical rate defining a frequent museumgoer (Hood, 1983). The high visitation rate ensured that the case would be active enough to generate the needed data.

Collecting and Analyzing Data Over Time

Case studies may be particularly effective in addressing the complexity presented by the context of time. The rich detail uncovered through a case study can reveal both macro-level learning mechanisms (e.g., how school curricula and museum visits can mutually support learning (Anderson, Lucas, Ginns, & Dierking, 2000) and the moment-by-moment interactions involved in learning in museums (Rowe, 2002). Whether macro or micro, the case study allows an exploration of why one particular interaction leads to another or how a learner's identity can be traced back through a developmental pathway of social interactions.

Systematic Analysis

Performing a case study requires that the rich data set be analyzed systematically, often using what is referred to as the constant comparative method within grounded theory (Glaser & Strauss, 1967). In this iterative method, each incident or data point gathered through interviews or other assessments is compared with all others. An explanatory theory soon emerges, which drives further data collection. Simultaneously, the researcher begins to generate categories or codes that group the incidents according to the theory. Once new data cease to add diversity to a category, the code is considered saturated. When all codes are saturated and the theory seems stable, writing begins. This entire process is similar to that undertaken in naturalistic inquiry.

Culturally Responsive Research and Evaluation

The diversity of cultures within the visiting (and potentially visiting) audience is one kind of variability that plays a foundational role in museum research and evaluation. By "culture" researchers mean a group's beliefs, attitudes, customs, values, ways of thinking, communication patterns, and frames of reference. While it often refers to ethnic or racial background, any group that has some shared affiliation and characteristics (e.g., people with disabilities, teenagers) can be seen as having a common culture.[2] In museum research and evaluation, many consider it vital to embrace this type of variability, both to accurately reflect the experiences of a wide range of visitors and to address the frequent calls to diversify museum audiences (e.g., American Association of Museums, 1992).

Culturally responsive approaches argue that no culture-free research or evaluation exists. For example, Ricardo Millett (2002) suggests that most evaluation instruments are developed for (and by) people who are employed, acculturated, well-educated, English-proficient, and enjoy moderate-to-high incomes. To this list could be added mobile, sighted, and hearing. In this view, part of the researcher's task is to actively cultivate one's own awareness and appreciation of cultural issues and their complexity, remaining vigilant to one's own potential cultural biases. Kirkhart (1995), among others, states that for an evaluation to have validity and utility, cultural perspectives must be addressed.

A culturally responsive framework affects all aspects of a study—from the formation of the research team to the dissemination of findings.

Inclusive Composition of the Research Team

In culturally responsive research and evaluation, some members of the research team are members of, or at least deeply understand, the specific cultures included in the study. This ensures that the research questions are appropriately framed and that the data collectors can recognize and interpret data appropriately. A variation of this type of inclusiveness is "participatory design," in which visitors are included as members of the design team rather than the research team, and are actively solicited for feedback during the creative process (Ringaert, 2001). Such a process may help to focus on the defects of the environment or exhibit, rather than the "defects" of the visitors, and is often used for research and evaluation relating to universal design (Gill, 1999).

Appropriate Design of Research Instruments

In culturally responsive research and evaluation, the instruments used to collect data are framed in a culturally appropriate fashion. Sometimes this involves minor revisions to a previously designed instrument to ensure that questions and scales are interpreted in similar ways across all cultural groups in the study ("cross-cultural equivalence"). Instruments can be developed in multiple languages using a "decentering translation" process, where an instrument is developed in one language, translated to the second language, then translated back to the original language to verify the quality of the original translation. Alternatively, existing methods and framings may be broadened to encompass a wider range of norms and values, such as redefining one's expectations of what "parental involvement" might look like for a particular cultural group (Garibay, 2006a).

In other situations, translations or adjustments may not adequately address cross-cultural differences, and more innovative methods may be useful. For example, storytelling may be an appropriate research technique for cultures rooted in oral traditions, and photographic methods may be more accurate for cultures that are not highly verbal. Participant observations, in which people share their thoughts with the researcher at times of their own choosing, may be more culturally appropriate for deaf visitors at hands-on exhibits than "think-aloud" protocols (Ericsson & Simon, 1993) where visitors talk and manipulate the exhibit simultaneously. This is because sign language, the primary language of the deaf culture, requires hand movements that can conflict with visitor manipulation of the exhibit (Reich, 2005).

Dissemination of the Findings to the Communities

Finally, adherents of culturally responsive research believe that disseminating data to the communities who participated in the research is important, and that the results of such studies should be easily accessible to respondents. Dissemination of results improves the validity of the findings by obtaining the community's input (e.g., Reich, 2000), and also serves as a political act by empowering community members with knowledge about themselves (Garibay, 2006b).

CONCLUSION

In this chapter, we presented two broad approaches for dealing with museum complexity in research and evaluation: reducing it to support cause-and-effect understanding, or embracing it to create a deep understanding of multiple perspectives. Pragmatic considerations such as constraints of time and budget make it virtually impossible to conduct the ideal versions of these approaches in real museums. But researchers who are familiar with a variety of techniques can make informed decisions about when to push for higher quality and greater understanding as resources permit. For example, evaluators working within an experimental paradigm may be able to include a small control group in their summative evaluations, and many formative evaluations can be adjusted to compare alternative offerings rather than just assessing visitors' experiences at one. Researchers wanting to conduct a more rigorous naturalistic inquiry can usually spend longer times with visitors to build more trust, and can always confirm with them at least the most important or controversial interpretations. Case studies can be scrutinized for the credibility of their arguments, not just the thickness of their descriptions. And researchers from *any* paradigm who want to incorporate culturally responsive frameworks, which are particularly time intensive, may be able to diversify their group of data collectors and adjust their methods to include the experiences of a previously excluded audience group.

While the two main approaches to complexity have been listed sequentially and separately in this chapter, several researchers and theorists (e.g., Guba & Lincoln, 1989; Patton, 2002) support the use of multiple methods to strengthen and triangulate interpretations of data, as long as the combination does not undermine any fundamental theoretical assumptions. For example, researchers who want to determine average holding times in an exhibition should use random sampling techniques when they track and time visitors. However, the results of such a study could be used in conjunction with a naturalistic study that purposively samples visitors for qualitative interviews. The union of these findings could yield an overview of typical traffic patterns and a range of the individual visitor experiences that motivate them. As another example, developers at the Museum of Science in Boston used universal design techniques to modify a diorama-based exhibition to be more accessible for all visitors, including those with disabilities. The summative evaluation (Davidson, 1991; Davidson, Heald, & Hein, 1991) included two main components: First, researchers conducted naturalistic interviews with persons with disabilities and found that the exhibit was accessible and engaging for that population. Next, researchers used random sampling and a pre/post experimental design with typical visitors, and concluded that holding time and conceptual understanding increased as a result of the exhibition modifications. The thoughtful and selective use of the two methodologies produced findings in support of the notion that universal design techniques may benefit all museum visitors.

As a final note, we encourage researchers and evaluators to discuss the values of these various approaches with stakeholders. Often museum practitioners will expect a certain study design because it is simple or familiar to them, but they may quickly embrace a different approach if they understand its purpose and efficacy. More importantly, stakeholders are the people making difficult decisions for action based on the completed studies, so it is important for them to appreciate the limitations of even the best work in social science. Sample sizes are always finite, generalizations are always tentative, and human behavior is always more complex than any single study can reveal.

ENDNOTES

This chapter benefited from the contributions and comments of Linda Deacon, Doreen Finkelstein, George Hein, the editors, and three anonymous reviewers.

1. This study used another slight modification from the method described, in that the versions of the exhibit rotated in sequence. This block design was a departure from a strict randomization, but was done because the interviews took over an hour and the research team wanted to ensure that every day each version was used at least once.

2. Despite the use of this term as a simplification, a cultural group is never homogenous; within-group diversity is often based on factors such as socioeconomic and educational background.

17

An Emerging Research Framework for Studying Free-Choice Learning and Schools

Laura M. W. Martin

Over the past ten years, discussion about informal learning, particularly in museums, has really taken off. The educational contributions of museums are being identified, along with questions of how museums support both learning and teaching. This chapter analyzes some of the current research on informal learning and schools as an example of how the field builds on ideas that were put forth at the conference *Public Institutions for Personal Learning* (Falk & Dierking, 1995), through the Museum Learning Collaboration (Schauble, Leinhardt, & Martin, 1997), and in other efforts (Hooper-Greenhill & Moussouri, 2002; Paris, 2002) of recent years. It draws on research related to learning in science centers, where most of the research has been conducted. It focuses on two key issues related to developments in the education field of the past decade: (1) integrating studies from multiple disciplines, and (2) characterizing properties of informal learning in museums. These two issues are related because our knowledge about the relationship between culture and thought with regard to museum learning is developing.

The first problem, that of integrating research perspectives from different fields, and the second, identifying characteristics of informal domains, also come together because the field of informal learning research touches on two murky areas: the interface between formal and informal institutional settings on a larger, societal level and the interface between intuitive and tutored thinking on the individual level. The institutional interface issue asks a question of how educational environments are arranged, and the individual interface issue asks a question about psychological development and how knowledge is constructed.

To address these, researchers look at a variety of features of *activity systems* that involve learning. That is, we focus on the active engagement of people in pursuits involving "socially organized domains of knowledge and technologies, including symbol systems" (Scribner, 1984). We ask: what does museum-going represent as a socially organized domain of activity?

If we are studying activity systems, both sets of questions mentioned above need to be addressed because we know that cultural practices and institutions (studied in the fields of education, sociology, and anthropology) shape the mind (studied in the field of psychology) and its habits (Rogoff, 2003a; Vygotsky, 1978). But we need to be clear that educational research sometimes balances questions of environment and development differently than psychological research does, and is often conducted differently. A useful framework for studying informal learning, then, encompasses both educational and psychological approaches in order to speak to the key unknowns simultaneously, the social and the individual.

INTEGRATING STUDIES FROM MULTIPLE DISCIPLINES

Let's look at what we know about learning from three traditions of research that have importance for understanding how museums can work with schools: cultural practice studies; specific content learning, in this case, science; and museum studies.

Cultural Practice Studies

In the past decade, learning research has expanded what we think about as factors influencing learning (see Falk & Dierking, 1995). A large body of evidence has convinced educational researchers and learning psychologists that cultural "noncognitive factors" (Cole, Griffin, & LCHC, 1987) are important ingredients in learning. Conversational patterns in one's home language (Rosebury, Warren, & Conant, 1992), the nature of participation structures in the home and elsewhere (Rogoff, 2003a), students' relations to the dominant culture (Ogbu, 1995), and their motivational states (Csikszentmihalyi & Hermanson, 1995) have been acknowledged as playing a role in learning and in how we view the learner. This research, which emphasizes culture and interaction in accounting for individual differences and patterns, can be loosely grouped under the heading of activity theory, discussed below.

Above all, we have learned that the cultural practices related to learning are key for both children and adults in determining how they internalize information and what kinds of thinking they generate as a result (Falk & Dierking, 2002; National Research Council, 2002). Research into family learning, workplace and everyday learning and problem solving, and work in community and after-school programs (Rogoff, 2003; see Falk & Dierking, 2002; Schauble & Glaser, 1996) points to the fact that the kind of thinking people engage in differs from place to place and from problem to problem (Beach, 1995). These studies note that interaction patterns differ and tools are used differently in different settings, with measurable outcomes in thinking and learning (Scribner & Cole, 1981).

Science Learning Research

In the past decade much also became known about the teaching and learning of science in schools. Science learning is viewed by many educational researchers as a special case of learning both because of the subject matter and because of how science learning has been viewed over time (Duschl, 1990; Lemke, 1993). Since science is thought of as having more and more impact and relevance to the general public (as opposed to a specialized public), questions of the best ways to introduce and learn about science are again being debated (see Chittenden, Farmelo, & Lewenstein, 2004). The need for public understanding has strengthened the debate about science learning outside the school walls, in the informal arena. Included in this debate are questions about what learning science means—a body of knowledge or a process or both (Duschl, 1994); psychological questions about how mental models develop (Schauble et al., 1991); and questions about how discourse forms affect conceptual learning (Ash, 2003; Duschl & Osborne, 2002).

When it comes to nonschool environments for learning science, less is known about how the instructional setting or people's everyday experiences with nature affect their knowledge, understanding, and interest in fundamental ways. We are not quite sure about the nature of everyday or intuitive science concepts arising from nonschool activities, nor are we quite clear about the value of informal experiences for developing science concepts. In other words, does what you learn at camp about nature help you in the classroom? We know that scientific discourse isn't likely to arise spontaneously on a casual field trip but we would like to know how cultural practices in informal educational settings can contribute to the development of such discourse and, in turn, to the development of a child's science concepts.

Museum-Learning Research

Studies of learning in science centers have contributed to what we know about informal science learning in general. The science museum-learning research, mostly looking at exhibit-related experiences, asks an array of questions that could be usefully linked in future research (Martin, 2001). Some of these relate to other kinds of museums as well. It points to the importance of:

1. *Structuring scientific knowledge.* Individuals can be shown to make conceptual gains from museum visits (Allen, 1997; Anderson, Lucas, & Ginns, 2003; Falk & Dierking, 1997; Stevens & Hall, 1997). These studies involve both unguided museum exploration and structured interventions. They are concerned with the content and reflexivity the visitor derives from a particular exhibit or exhibition.
2. *The role of social interaction.* Family members engage in a variety of strategies that encourage explanation and understanding. This research views the group as the learning unit and examines the nature of group interactions to either infer about or correlate with evidence of learning (e.g., conceptual gain; information acquired; depth of understanding; Borun, Chambers, & Cleghorn, 1996; Crowley, Callanan, Tenebaum, & Allen, 2001; Crowley & Jacobs, 2002; Ellenbogen, 2002; Moussouri, 1997).
3. *Mediating experiences and devices.* The organization of students' class visits to museums can relate to the depth and retention of visit content (Anderson & Lucas, 2003). Studies of field trips emphasize activities and materials that enhance school visits as measured by

the richness of children's concept mapping or postvisit discussions (Anderson, Lucas, Ginns, & Dierking, 2000). Evidence of learning and thinking has also been detected through social- and object-based mediation among members of museum communities as well as among visitors (vom Lehn, Heath, & Hindmarsh, 2001); studies from the Museum Learning Collaboration looked at a variety of learners, including docents and visitors, and found evidence of motivation and knowledge gain (Abu-Shumays & Leinhardt, 2002; Leinhardt, Knutsen, & Crowley, 2003).

4. *Institutional meaning.* Educational theorists (Bruner, 1996; Ogbu, 1995) and others have argued that an individual's identity as a learner is shaped by the cultural institutions they come in contact with as well as by day-to-day moments in life. There is evidence that formal institutions of learning are not cultivating many people's identities as learners of science and are even systematically discouraging that development for certain groups. There is simultaneously an acknowledgment that informal settings like museums may do a better job building learner identity for some.

In addition to describing learning activity, museum studies demonstrate that learning from museum visits is detectable and can be manipulated. They point to the fact that, depending on the form of the exhibit, the nature of access to the exhibit, the social support at the exhibit and before and after visiting, the relationships between group members, and the purpose of the visit, different forms of interactions occur and different degrees of attention, encoding, construction, and recall of science concepts can be measured.

Overall, the picture from these three areas is not entirely coherent. Studies encompass educational research and practice, social science, and program design. Studies focus on schools, museums and other informal situations, their programs, participants, and their material aspects. Still, the ensemble is a valuable, multifaceted effort (Lewis & Kelemen, 2002) that responds to the current context of our knowledge about museum learning where a confluence of pedagogical and theoretical questions has resulted from traditions of research on out-of-school learning contexts and of design experiments in museums.

CHARACTERIZING PROPERTIES OF TEACHING AND LEARNING

The second issue—that of identifying characteristics of informal learning in museums—leads us to consult social science literature that has looked at traditional cultures and other nonschool environments.

Social science has looked at out-of-school learning for decades and as a result, the dynamics of cognitive functioning in social contexts have been described and have even been used to inform current school research (e.g., Lemke, 1993). These findings are a good starting point for looking at out-of-school experiences in our own society.

Most of the work on out-of-school learning systems has been done in traditional societies with the disciplinary tools of cognitive anthropology and cultural psychology (Cole, 1996; Hutchins, 1995; Rogoff, 2003a). It focuses on the context, tools, social relations, as well as the assumptions underlying cultural practices of teaching and learning. The related body of work has characterized learning activity in the following ways, as expressed by Scribner and Cole (1973) and summarized in table 17.1.

Scribner and Cole do not find a clear picture of "noninstitutional formal education." According to their distinctions, museum learning may be said to straddle the "formal" and "informal," so that if we were to add a column to the table representing museum learning, we would have to perhaps combine definitions from the other columns. Because the research on mu-

Table 17.1. Characteristics of Learning Activity In and Out of School

Informal Learning	*Learning in Westernized School Settings*
Occurs in the course of mundane adult activities in which the young take part according to their abilities	Emphasizes universalistic values, criteria, and standards of performance
Occurs in families; expectations are in terms of who a person is not what was accomplished	What is being taught more important than who is doing the teaching
Fosters traditionalism	May represent a culture that denigrates the indigenous culture
Fuses emotional and intellectual domains	Emphasizes language; language occurs out of context
Is strongly observational, participatory	Emphasizes mastering symbol systems
Occurs where meaning is intrinsic to context	Introduces new subjects, unknown history, and physical universe not derived from senses

Terms adapted from Scribner & Cole, 1973.

seum learning has taken different disciplinary approaches, however, we are not at the point where we can definitively identify its unique characteristics. Some approaches have been attempted, for instance by Hein (1998) and Falk and Dierking (2002), but it can be argued that this work categorizes theories rather than integrates findings from different types of studies.

Another set of distinctions in social science literature looks at the nature of mental activity in different settings. If we compare what the literature characterizes as features of thinking in and out of school, we infer that there are fundamental differences between the learning systems there, specifically in the nature of the tasks, participant structures, and physical settings that give rise to deeper understanding and scientific reasoning.

For example, school thinking and practical thinking (that is, thinking that happens informally in the workplace or in the course of everyday tasks) can be abstractly characterized, as shown in table 17.2.

According to this viewpoint, thinking (and learning) unfold differently in various institutional settings because the activity is different. Furthermore, the nature of the goal formed by the actors and their resulting discourse and tool use may lead to different patterns of internalization processes.

Thinking at the interface between two learning systems, where museums and schools may want to work together, has not yet been captured. Of course, the problem of defining "informal" in a new way remains but, at this point, it is fair to say that most researchers are uncomfortable with a distinction between activities based solely on their location (Rogoff, 2003b; Martin, 2001).

Table 17.2. Characteristics of Thinking In and Out of Schools

Thinking Taught in Schools	*Practical Thinking*
Generalized	Continually creative
No nonsymbolic content	Strives for mastery of the concrete
Emphasizes mental processes	Utilizes one's knowledge
Verbal	Uses whatever works
Solitary	Involves others
Independent of a specific end	Sensitive to the environment
Uses general tools	Uses available materials
Built on many systematic examples	Flexible and economical

Terms adapted from Scribner, 1985.

AN EMERGING FRAMEWORK FOR EXAMINING INFORMAL LEARNING AND SCHOOLS

Looking at learning activity as a system derives from activity theory, a variant of what is variously termed cultural historical or sociocultural psychology (Cole, 1996; Scribner, 1984; Wertsch, 1985; Zinchenko & Kozulin, 1986). The advantage of adopting a sociocultural framework is twofold: first, because it assumes the cultural formation of mind and so allows us to focus at different levels of learning practices, for example, on moments of interaction or on institutional developments; and second, because it views learning as a cultural practice and takes into consideration the mediation of tool use, language use, participation structures, and social practices, a sociocultural approach allows comparisons of learning activity or learning systems between settings. Importantly, it can account for different learning outcomes in different settings (Laboratory of Comparative Human Cognition, 1983), which traditional cognitive psychology could not. For example, it allows us to see how familial discourse is brought to school and how a school may or may not leverage that for instruction (Rosebery, Warren, & Conant, 1989).

The sociocultural framework does not dictate exact units of analysis. Some of the researchers working on these problems are looking at features of participant structures, others at discourse and argumentation forms, others at physical behaviors and patterns of tool use. Still others look at institutional dynamics that have an impact on choices made by individuals carrying out practices in the institutions. What we know about the psychology of learning and of the science of studying learning is that, depending on the goals of the actors and the tools at hand (symbolic, physical, psychological), the unit of activity to be analyzed might need to change. So, depending on the question, one might look at the activity associated with solving a single problem or at the activity associated with a semester-long curriculum unit. Furthermore, we know the mental operations an individual brings to bear on a problem may shift according to the problem's context, for example, school or workplace (Istomina, 1975; Lave, 1993; Scribner, 1990). Multiple goals and multiple voices or discourses can be at work simultaneously in the meaning-making process. Sociocultural theory accommodates this multiplicity.

The theory does *not* help us with the fact that most of our thinking about thinking is based on Western, formal educational practices and settings: right answers, good grades, long-term recall, and so on. This means it is very hard to think about what learning looks like and what transmission of cultural knowledge looks like when it concerns curricular areas outside of school, since we tend to treat right answers and test scores as the only "scientific" evidence of learning. Creative analyses need to accrue here, and that body of work needs to be stitched together.

Although a multifaceted research approach encompasses science education, experimental developmental psychology, sociology, and program evaluation, this does not mean that our observations are simultaneously true and not-true in a postmodern, relativist way (Lewis & Kelemen, 2002). Nor does it mean that any and all analytic tools (for example, constructivism, brain research, and Piagetian psychology) are useful. Rather, a multifaceted work plan based on sociocultural approaches will best capture the features of the complex learning systems we are studying. Multiple research voices within a project are needed, juxtaposing studies of identified epistemologies with naturalistic studies, individual factors with institutional factors, and using them to derive a useful characterization of the opportunities offered by informal museum learning.

IDENTIFYING CRITICAL PROPERTIES OF INFORMAL LEARNING AND THEIR RELATION TO SCIENCE LEARNING

If we assume that setting, activity and goal, mediators, and participation structures affect learning (Falk & Dierking, 2002), we need to identify what is happening with them in "informal" or nonschool contexts. Because we are interested in the *psychology of learning*, some researchers are starting with what cultural psychology says about how people engage each other in activity. The researchers working from observational traditions of developmental psychology take an inductive approach. They are looking at what unfolds in informal settings with an eye to improving learning for diverse individuals. The assumption is that if we look at what people do in different settings, we may identify patterns of cultural transmission that may be deliberate or incidental and through which intended and unintended meaning-making may be the outcome.

Because of an interest in the *improvement of teaching and learning* as well, researchers also employ the traditions of educational research in order to ultimately incorporate nonschool learning elements into professional development workshops in science for teachers and into studying teacher enactments in practice. Through education research, researchers work with analytic tools that arise from the rational tradition of the logic and sequence of scientific thinking and explanation evidenced in participants' verbalizations, actions, and other responses.

Science education research does not need to make claims about the reality of abstract relationships between objects in nature that the learner may be attempting to understand, but it does claim that Western civilization has evolved a canon of scientific concepts and processes that are the basis for what and how we teach, and upon which we reflect. Even as this canon changes with discoveries and breakthroughs in the sciences, the content of science is, ideally, accessible by people in our culture, should they learn its "language." These science education studies, therefore, look for scientific reasoning among learners and for the development of higher-order representation and canonical mental modeling among them. The research examines argumentation, conversation, and science process skills among participants in learning activities. Rather than focus solely on the mind of the individual learner, however, researchers working with this approach look at interindividual learning processes as well, and trace the origins of learning to the tasks, interaction patterns, and affordances of the environment. These might include access to various cognitive tools like learning conversations or new collaborative learning technologies. At a systems level, researchers will also need to understand the larger activity systems, policies, and institutional practices that enable connections between informal and formal systems to occur. Multiple probes and methodologies are appropriate; the domain is complex and requires a number of investigative tools to proceed to illuminate the most promising structures and mechanisms of science learning at the intersection of informal and formal learning.

In imagining a new research agenda, three areas emerge as promising ones for further investigation of learning in informal settings and schools: participation, explanation, and design for learning. These three areas raise questions, accessible to both researchers and practitioners, that could link both empirical studies of individuals, groups, and institutions to theoretically based studies of informal learning and practical interventions in the field.

Questions About Participation

A key research question here asks how learning activity from the informal arena can connect to the classroom. Do activity and discussion patterns among different families and families of different heritages affect participation by children in school science?[1]

Questions about participation also concern institutions and their organizational structures which may permit particular forms of participation to occur. We know this from work in the area of school and business organization theory. How are schools and museums organized and structured such that they can operate in fruitful collaboration with each other? What does the learning environment of museums mean in our culture and how do museums incorporate participation in canonical science learning into their practices? What do teachers with different degrees of involvement in science centers understand about museums as science resources? What are the standards by which museums judge their work with schools?

Questions About Explanation, Argumentation, and Discourse

Explanations and explanatory processes are not a new research area but represent a research focus that has critical importance and potential impact across multiple settings as learning and facilitation tools themselves. How do museum conversations relate to children's developing understandings of science concepts, particularly conversation prompted by different exhibit forms and their relation to children's symbolic understanding of symbols? How do museum visitors engage in deliberative and argumentative discourse around exhibits such that they come to understand how science is carried out? How is scientific thinking supported and mediated by informal educators? What are the variations of and supports for scientific reasoning and explanation in informal settings?

Questions About Designs for Learning

Questions about the designs for learning explore ways in which interventions can be created to effect productive change in learning environments and educational settings. How might teachers' scientific reasoning develop during a museum staff development workshop designed to change inquiry practices? Because the museum environment is nonacademic, does teachers' thinking itself unfold differently than in a traditional university or district in-service program? And how can we design probes to measure this? Does this make a difference in their teaching practices? The notion of designing for learning also addresses, for example, how the physical forms of different exhibits can be designed—as models, metaphors, push-buttons, and so forth—that affect participation in talk and concept development.

Studies addressing these questions will begin to fill in the picture of what characterizes informal science learning and its relationship to schools, and to the science education field as a whole.

CONCLUSION

Since 1994 several important research strands and traditions have been converging on the problem of studying science learning outside of classrooms or culturally "formal" practices. We are at the point where the questions and approaches are becoming clear and where they can be better integrated. In the shared arena of schools and museums, many different kinds of collaborative projects are being implemented, funded, and evaluated, as are different research initiatives. Discourse in this arena is not coherent yet—which is a healthy sign—but it is advancing. Using a framework that encompasses critical methodologies from subject matter education, cognitive development, and cultural psychology within a sociocultural framework, it seems diverse threads can be tied together to address fundamental questions about how learning happens in informal settings so that learning in schools—specifically, science learning—can be successfully supported by museums.

Current research work encompasses multiple approaches to studying informal learning in order to speak to the complex nature of many of the basic questions we have. Previous work in social science has provided a basis on which to move forward and identify how learning comes about in the contexts of interest to us. Work in education, psychology, and museum studies has informed the discussion substantively over the past decade.

At this point in our understanding, we need analytic tools that consider learning as a cultural practice in order to integrate previous work and to move the work forward. These tools do not limit the questions we can ask, rather, they allow us to create a fuller, valid picture of a complex process at work at societal and individual levels. Some of us are interested in studying the practices that communicate cultural meaning; some, the meaning people take away with them from encounters with artifacts and with other people; and some, the mechanisms by which museum visitors construct identity and meaning. Particular issues that may be in need of further examination are: the relation of home patterns of learning conversations to what visitors bring to a museum; of learning interaction patterns of schools and of what teachers and students bring to museum visits; how the informal setting might support meaning-making in science for people with diverse cultural practices; and, how scientific reasoning can develop through the kinds of experiences informal settings provide.

As we work to demonstrate or justify the educational role of museums, not just understand its basic dynamics, we help identify bridges between the worlds of home and school and we help unpackage the effect and power of informal settings on children's intellectual development. The task of researchers in this field is to develop a common basis for discussion, identify critical research issues, and prepare students and museum educators to study and design effective learning. Through this work and through continuing dialogue among researchers, discussion can be stimulated across institutions and beyond specific projects into the next decade.

ENDNOTES

This work was supported by a grant from NSF (No. ESI–019787) to the Exploratorium. An earlier version of this chapter was published in *Science Education*, 88 (suppl.1): S34–S47, 2004, © Wiley Periodicals, Inc. Used with permission of John Wiley & Sons, Inc.

1. The notion of an intent participation structures, developed by Rogoff et al. (2003) is one that provides a template for comparing teaching strategies in different contexts. This template is likely to prove useful in comparing features of different examples of learning activity.

References

Abraham, M., Griffin, D. J. G., & Crawford, J. (1999). Organization change and management decision in museums. *Management Decision, 37*(10), 736–751.

Abu-Shumays, M., & Leinhardt, G. (2002). Two docents in three museums: Central and peripheral participation. In G. Leinhardt, K. Crowley, & K. Knutson (Eds.), *Learning conversations in museums* (pp. 45–80). Mahwah, NJ: Lawrence Erlbaum.

Adelman, L. M., Falk, J. H., & James, S. (2000). Assessing the National Aquarium in Baltimore's impact on visitor's conservation knowledge, attitudes and behaviors. *Curator, 43*(1), 33–62.

Allen, S. (1997). Using scientific inquiry activity in exhibit explanations. *Science Education, 81*(6), 715–734.

Allen, S. (2002). Looking for learning in visitor talk: A methodological exploration. In G. Leinhardt, K. Crowley, & K. Knutson (Eds.), *Learning conversations in museums* (pp. 259–303). Mahwah, NJ: Lawrence Erlbaum.

Allen, S. (2003, October). *To partition or not to partition: The impact of walls on visitor behavior at an exhibit cluster.* Paper presented at the meeting of the Association of Science-Technology Centers, Minneapolis, MN.

Allen, S. (2004). *Finding significance.* San Francisco, CA: Exploratorium.

Allen, S., & Feinstein, N. (2006). The effect of physical interactivity on visitor behavior and learning. Manuscript in preparation.

Allen, S., & Gutwill, J. (2004). Designing with multiple interactives: Five common pitfalls. *Curator, 47*(2), 199–212.

Alpers, S. (1991). The museum as a way of seeing. In I. Karp and S. Levine (Eds.), *Exhibiting cultures: The poetics and politics of museum display* (pp. 25–42). Washington, DC: Smithsonian Institution Press.

American Association of Museums. (1992). *Excellence and equity: Education and the public dimension of museums.* Washington, DC: American Association of Museums.

Anderson, D. (1999a). *The development of science concepts emergent from science museum and post-visit activity experiences: Students' construction of knowledge.* Unpublished doctorial dissertation, Queensland University of Technology, Brisbane, Australia.

Anderson, D. (1999b). *Understanding the impact of post-visit activities on students' knowledge construction of electricity and magnetism as a result of a visit to an interactive science centre.* Brisbane, Australia: Queensland University of Technology.

Anderson, D. (2003). Visitor's long-term memories of world expositions. *Curator, 46*(4), 400–420.

Anderson, D. (2004). *Reluctant learners: Art museums in the twenty-first century.* Helsinki: National Museum of Art.

Anderson, D., Bibas, D., Brickhouse, N., Brown, J., Dierking, L. D., Ellenbogen, K. M., Falk, J. H., Klotz, S., Rounds, J., Semmel, M., Serrell, B., Spock, M, & Stein, J. (2006). Long-term impacts. In J. Stein, L. D. Dierking, J. H. Falk, & K. M. Ellenbogen (Eds.), *Insights: A museum learning resource.* Available from the Institute for Learning Innovation, 166 West Street, Annapolis, MD 21401. Online at www.ilinet.org/ipip/In_Principle_In_Practice_Insights_-_Museum_Learning_Resource.pdf.

Anderson, D., Kisiel, J., & Storksdieck, M. (2006). Understanding teachers' perspectives on field trips: Discovering common ground in three countries. *Curator, 49*(3), 365–386.

Anderson, D., & Lucas, K. B. (2001). A wider perspective on museum learning: Principles for effective post-visit activities for enhancing student learning. In S. Errington, S. Stocklmayer, & B. Honeyman (Eds.), *Using museums to popularise science and technology* (pp. 131–141). London: Commonwealth Secretariat.

Anderson, D., Lucas, K. B., & Ginns, I. S. (2003). Theoretical perspectives on learning in an informal setting. *Journal of Research in Science Teaching, 40*(2), 177–199.

Anderson, D., Lucas, K. B., Ginns, I. S., & Dierking, L. D. (2000). Development of knowledge about electricity and magnetism during a visit to a science museum and related post-visit activities. *Science Education, 84*(5), 658–679.

Anderson, D., & Piscitelli, B. (2002). Parental recollections of childhood museum visits. *Museum National, 10*(4), 26–27.

Anderson, D., Piscitelli, B., Weier, K., Everett, M., & Tayler, C. (2002). Children's museum experiences: Identifying powerful mediators of learning. *Curator, 45*(3), 213–231.

Anderson, D., & Shimizu, H. (2007). Factors shaping vividness of memory episodes: Visitor's long-term memories of 1970 Japan World Exposition. *Memory, 15*(2), 177–191.

Anderson, G., & Arsenault, N. (2004). *Fundamentals of educational research* (2nd ed.). London: Routledge.

Anderson, L. (2005). *Chicago Tribune*. Available at www.centredaily.com, Saturday, August 13, 2005.

Anonymous. (2005, December 24). Jesus, CEO. *The Economist, 377*(8458), 41–44.

Ansbacher, T. (2002a). On making exhibits engaging and interesting. *Curator, 45*(3), 167–173.

Ansbacher, T. (2002b, March–April). What are we learning? Outcomes of the museum experience. *Informal Learning Review 53*.

Ash, D. (2003). Dialogic inquiry in life science conversations of family groups in a museum. *Journal of Research in Science Teaching, 40*(2), 138–162.

Bailey, E., Bronnenkant, K., Kelley, J., & Hein, G. (1998). Visitor behavior at a constructivist exhibition: Evaluating *Investigate* at Boston's Museum of Science. In C. Dufresne-Tasse (Ed.), *Evaluation et education museal: Nouvelles tendances* (pp. 149–168). Montreal: ICOM/CECA.

Bakeman, R., & Gottman, J. M. (1997). *Observing interaction: An introduction to sequential analysis.* Cambridge: Cambridge University Press.

Ballantyne, R., Fien, J., & Packer, J. (2001). School environmental education programme impacts upon student and family learning: A case study analysis. *Environmental Education Research, 7*(2), 23–37.

Ballantyne, R., & Packer, J. (2002). Nature-based excursions: School students' perceptions of learning in natural environments. *International Research in Geographical and Environmental Education, 11*(3), 11.

Ballantyne, R., & Packer, J. (2005). Promoting environmentally sustainable attitudes and behaviour through free-choice learning experiences: What is the state of the game? *Environmental Education Research, 11*(3), 281–296.

Balling, J. D., Falk, J., & Aronson, R. (1992). *Pre-trip orientations: An exploration of their effects on learning from a single visit field trip to a zoological park.* Final report, NSF Grant SED77–18913.

Bamberger, Y., & Tal, T. (2005). *Learning in a personal-context: Levels of choice in a free-choice learning environment at science and natural history museums.* Paper presented at the European Association for Research on Learning and Instruction conference, Nicosia, Cyprus.

Barab, S. A., & Kirshner, D. (2001). Rethinking methodology in the learning sciences. *Journal of the Learning Sciences, 10*(1&2), 5–15.

Barriault, C. (1999). The science center learning experience: A visitor-based framework. *Informal Learning Review, 35*(1), 14–16.

Barton, A. C., Drake, C., Perez, J. G., St. Louis, K., & George, M. (2004). Ecologies of parental engagement in urban education. *Educational Researcher*, *33*(44), 3–12.

Barton, J., & Kindberg, T. (2001). *The challenges and opportunities of integrating the physical world and networked systems.* HPL Technical Report HPL-2001–18.

Baxandall, M. (1987). *Patterns of intention: On the historical explanation of pictures.* New Haven, CT: Yale University Press.

Beach, K. (1995). Sociocultural change, activity, and individual development: Some methodological aspects. *Mind, Culture and Activity: An International Journal*, *2*, 277–284.

Bedford, L. (2001). Storytelling: The real work of museums. *Curator*, *44*(1), 27–34.

Beetlestone, J. G., Johnson, C. H., Quin, M., & White, H. (1998). The science center movement: Contexts, practice, next challenges. *Public Understanding of Science*, *7*(1), 5–26.

Bell, L. (2000, July–August). Assuring intellectual access: Lessons from Boston's Museum of Science. *Dimensions*, 3–5.

Bennett, T. (2004). *Pasts beyond memory: Evolution, museums, colonialism.* New York: Routledge.

Bentley, D., & Watts, M. (1993). *Learning and teaching in school science: Practical alternatives.* Philadelphia: Open University Press.

Benton Foundation, The. (2001). *Partners in public service: Models for collaboration.* University Park: Penn State Public Broadcasting at Pennsylvania State University.

Bielick, S., & Karns, D. (1998). *Still thinking about thinking: A 1997 telephone follow-up study of visitors to the Think Tank exhibition at the National Zoological Park.* Washington, DC: Smithsonian Institution, Institutional Studies Office.

Birken, L. (1992). What is Western civilization? *History Teacher*, *25*(4), 451–461.

Bitgood, S., Serrell, B., & Thompson, D. (1994). The impact of informal education on visitors to museums. In V. Crane et al. (Eds.), *Informal Science Learning* (pp. 61–106). Dedham, MA: Research Communications.

Bogner, F. X. (1998). The influence of short-term outdoor ecology education on long-term variables of environmental perspectives. *Journal of Environmental Education*, *29*(4), 17–29.

Borun, M., Chambers, M. B., & Cleghorn, A. (1996). Families are learning in science museums. *Curator*, *39*(2), 124–138.

Borun, M., Chambers, M. B., Dritsas, J., & Johnson, J. I. (1997). Enhancing family learning through exhibits. *Curator*, *40*(4), 279–295.

Borun, M., & Dritsas, J. (1997). Developing family-friendly exhibits. *Curator*, *40*(3), 178–196.

Borun, M., Dritsas, J., Johnson, J. I., Peter, N. E., Wagner, K. F., Fadigan, K., Jangaard, A., Stroup, E., & Wenger, A. (1998). *Family learning in museums: The PISEC perspective.* Philadelphia, PA: Franklin Institute.

Bowen, J. P., & Filippini-Fantoni, S. (2004). *Personalization and the web from a museum perspective.* Paper presented at the Museums and the Web annual conference. Retrieved January 20, 2005, from *Papers: Museums and the Web 2004*, www.archimuse.com/mw2004/papers/bowen/bowen.html.

Bradburne, J. (1998). Dinosaurs and white elephants: The science centre in the twentieth century. *Public Understanding of Science*, *7*, 237–253.

Bradburne, J. M. (2004). The museum time bomb: Overbuilt, overtraded, overdrawn. *Informal Learning Review*, *65*(1), 4–13.

Bransford, J. D., Brown, A. L., & Cocking, R. (1999). *How people learn: Brain, mind, experience and school.* Washington, DC: National Research Council.

Breidenbach, J., & Zukrigl, I. (1999). The dynamics of cultural globalization: The myths of cultural globalization. *Cultural Collaboratory.* Retrieved May 13, 2006, from www.inst.at/studies/collab/breidenb.htm.

Briseno, A. (2005). *Adult learning experiences from an aquarium visit: The on-site and longitudinal role of personal agendas and social interactions in family groups.* Unpublished master's thesis, University of British Columbia, Vancouver, Canada.

Brown, A. L. (1992). Design experiments: Theoretical and methodological challenges in creating complex interventions in classroom settings. *Journal of the Learning Sciences*, *2*, 141–178.

Brown, J. S., Collins, A., & Duguid, P. (1989). Situated cognition and the culture of learning. *Educational Researcher*, *18*(1), 32–42.

Bruner, J. (1986). *Actual minds, possible worlds.* Cambridge, MA: Harvard University Press.

Bruner, J. (1994). The "remembered" self. In U. Neisser (Ed.), *The remembering self: Construction and accuracy in self-narratives* (pp. 41–54). New York: Cambridge University Press.

Bruner, J. (1996). *The culture of education.* Cambridge, MA: Harvard University Press.

Campbell, D. T., & Fiske, D. W. (1959). Convergent and discriminant validation by the multitrait-multimethod matrix. *Psychological Bulletin*, *56*, 81–105.

Campbell, D. T., & Stanley, J. C. (1963). *Experimental and quasi-experimental designs for research.* Boston: Houghton Mifflin.

Cannon-Brookes, P. (1997). The presentation of contemporary science and technology in museums and science centers. *Museum Management and Curatorship, 16*(2), 201–203.

Carr, D. (2003). *The promise of cultural institutions.* Walnut Creek, CA: AltaMira Press.

Chesebrough, D. E. (2005). New models: The search for an improved approach to science centers and museums. *The Informal Learning Review, 74*(September–October), 1, 8–13.

Chew, R. (2003). Community roots. In American Association of Museums (Ed.), *Mastering civic engagement: A challenge to museums* (pp. 63–64). Washington, DC: American Association of Museums.

Chicago Tribune. (2005). *Museum of Earth History.* Retrieved August 17, 2005, from www.centredaily.com.

Children's Voices, Insights, 2005. Retrieved May 30, 2006, from www.ilinet .org/ipip/In_Principle_In_Practice_Insights_-_A_Museum_Learning_Resource.pdf.

Chittenden, D., Farmelo, G., & Lewenstein, B. V. (Eds.). (2004). *Creating connections: Museums and the public understanding of current research.* San Francisco: AltaMira Press.

Cobb, P., Confrey, J., diSessa, A., Lehrer, R., & Schauble, L. (2003). Design experiments in educational research. *Educational Researcher, 32*(1), 9–13.

Coglisera, C. C., & Brighamb, K. H. (2004). The intersection of leadership and entrepreneurship: Mutual lessons to be learned. *Leadership Quarterly, 15*(6), 771–799.

Cole, M. (1996). *Cultural psychology: A once and future discipline.* Cambridge, MA: Belknap Press.

Cole, M., Griffin, P., & LCHC. (1987). *Contextual factors in education.* Madison: Wisconsin Center for Education Research.

Collaborative Digitization Program. (n.d.). Collaborative Digitization Program's website. Retrieved from www.cdpheritage.org/index.cfm.

Collins, A. (1999). The changing infrastructure of education research. In E. C. Lagemann & L. S. Shulman (Eds.), *Issues in education research: Problems and possibilities* (pp. 289–298). San Francisco: Jossey-Bass.

Collins, J. (2001). *Good to great: Why some companies make the leap . . . and others don't.* New York: HarperCollins.

Collins, J. (2005). *Good to great and the social sectors: A monograph to accompany good to great.* Boulder, CO: Jim Collins.

Collins, J. C., & Porras, J. I. (1994). *Built to last. Successful habits of visionary companies.* London: Century.

Commission on Museums for a New Century. (1984). *Museums for a new century.* Washington, DC: American Association of Museums.

Cook, T. D., & Campbell, D. T. (1979). *Quasi-experimentation: Design and analysis issues for field settings.* Boston: Houghton Mifflin.

Cooks, R. (1999). Is there a way to make controversial exhibits that work? *Journal of Museum Education, 23*(3), 18–20.

Correa-Chávez, M., Rogoff, B., & Mejia, R. (2005). Cultural patterns in attending to two events at once. *Child Development, 76*(3), 664–678.

Crowley, K., Callanan, M. A., Jipson, J., Galco, J., Topping, K., & Shrager, J. (2001). Shared scientific thinking in everyday parent-child activity. *Science Education, 85*(6), 712–732.

Crowley, K., Callanan, M. A., Tenenbaum, H. R., & Allen, E. (2001). Parents explain more often to boys than to girls during shared scientific thinking. *Psychological Science, 12*(3), 258–261.

Crowley, K., & Jacobs, M. (2002). Building islands of expertise in everyday family activity. In G. Leinhardt, K. Crowley, & K. Knutson (Eds.), *Learning conversations in museums* (pp. 259–303). Mahwah, NJ: Lawrence Erlbaum Associates.

Csikszentmihalyi, M., & Hermanson, K. (1995). Intrinsic motivation in museums: Why does one want to learn? In J. H. Falk & L. D. Dierking (Eds.), *Public institutions for personal learning: Establishing a research agenda* (pp. 67–78). Washington, DC: American Association of Museums.

Damasio, A. R. (1994). *Descartes' error: Emotion, reasons, and the human brain.* New York: Avon Books.

Dana, J. C. (1917). *The new museum: Selected writings.* Washington, DC: American Association of Museums. (Reprinted in 1999)

Davidson, B. (1991). *New dimensions for traditional dioramas, multisensory additions for access, interest and learning.* Boston: National Science Foundation and Boston Museum of Science.

Davidson, B. (2001). *Universal Design (Accessibility).* Retrieved December 30, 2003, from www.mos.org/exhibitdevelopment/access.

Davidson, B., Heald, C. L., & Hein, G. (1991). Increased exhibit accessibility through multisensory interaction. *Curator, 34*(4), 273–290.

Davis, J., Gurian, E. H., & Koster, E. H. (2003). Timeliness: A discussion for museums. *Curator, 46*(4), 353–361.

Dean, D., & Rider, P. E. (2005). Museums, nation and political history in the Australian National Museum and the Canadian Museum of Civilization. *Museum and Society, 3*(1), 35–50.

Deci, E. L., & Ryan, R. M. (1985). *Intrinsic motivation and self-determination in human behavior.* New York: Plenum.

Dettmann-Easler, D., & Pease, J. L. (1999). Evaluating the effectiveness of residential environmental education programs in fostering positive attitudes toward wildlife. *Journal of Environmental Education, 31*(1), 33–39.

Dewey, J. (1938). *Experience and education.* New York: Collier Books, Macmillan.

DeWitts, J. & Osborne, J. (in press). Supporting teachers on science-focused school trips: towards an integrated framework of theory and practice. *International Journal of Science Education.*

Diamond, J. (1986). The behavior of family groups in science museums. *Curator, 29*(2), 139–154.

Dierking, L. D. (1989). What research says to museum educators about the family museum experience. *Journal of Museum Education, 14*(2), 9–11.

Dierking, L. D. (in press). Museums, affect & cognition: The view from another window. In S. Alsop (Ed.), *The affective dimensions of cognition: Studies from education in the sciences.* London: Kluwer Press.

Dierking, L. D., Andersen, N., Ellenbogen, K. M., Donnelly, C., Luke, J. J., & Cunningham, K. (2005, Spring). The Family Learning Initiative at The Children's Museum of Indianapolis: Integrating research, practice & assessment. *Hand-to-Hand.*

Dierking, L. D., Cohen Jones, M., Wadman, M., Falk, J. H., Storksdieck, M., & Ellenbogen, K. M. (2002, July–August). Broadening our notions of the impact of free-choice learning experiences. *Informal Learning Review, 55*, 1, 4–7.

Dierking, L. D., Ellenbogen, K. M., & Falk, J. H. (Eds.). (2004). In principle, in practice: Perspectives on a decade of museum learning research (1994–2004). *Science Education, 88*, Supp. 1.

Dierking, L. D., & Falk, J. H. (1994). Family behavior and learning in informal science settings: A review of the research. *Science Education, 78*(1), 57–72.

Dierking, L. D., & Falk, J. H. (2003). Optimizing out-of-school time: The role of free-choice learning. *New Directions for Youth Development, 97*, 75–89.

Dierking, L. D., Falk, J. H., Rennie, L., Anderson, D., & Ellenbogen, K. (2003). Policy statement of the "Informal Science Education" Ad Hoc Committee. *Journal of Research in Science Teaching, 40*, 108–111.

Dierking, L. D., Luke, J., Foat, K., & Adelman, L. (2000). Families and free-choice learning. *Museum News, 80*(6), 38–43, 67.

Dierking, L. D., & Pollock, W. (1998). *Questioning assumptions: An introduction to front-end studies in museums.* Washington, DC: Association of Science-Technology Centers.

Dierking, L. D., Rand, A. G., Solvay, M., & Kiihne, R. (2006). Laughing & learning together: Family learning research becomes practice at the USS Constitution Museum. *History News, 61*(3), 12–15.

Doering, Z. D., & Pekarik, A. J. (1996). Questioning the entrance narrative. *Journal of Museum Education, 21*(3), 20–22.

Dreyfus, H. (1967). *Designing for people.* New York: Paragraphic Books.

Driver, R., Asoko, H., Leach, J., Mortimer, E., & Scott, P. (1994). Constructing scientific knowledge in the classroom. *Educational Researcher, 23*(7), 5–11.

Durant, J. (1996). Introduction. In J. Durant (Ed.), *Museums and the public understanding of science* (pp. 7–11). London: Science Museum & Committee on the Public Understanding of Science.

Duschl, R. (1990). *Restructuring science education: The importance of theories and their development.* New York: Teachers College Press.

Duschl, R. (1994). Research on the history and philosophy of science. In D. L. Gabel (Ed.), *Handbook of research on science teaching and learning* (pp. 443–465). New York: Macmillan.

Duschl, R. A., & Osborne, J. (2002). Supporting and promoting argumentation discourse. *Studies in Science Education, 38*, 39–72.

Dweck, C. S. (1989). Motivation. In A. Lesgold & R. Glaser (Eds.), *Foundation for a psychology of education* (pp. 87–136). Hillsdale, NJ: Lawrence Erlbaum Associates.

Edelman, G. (1987). *Neural Darwinism: The theory of group selection.* New York: Basic Books.

Electronic Guidebook Forum. (2001). Lessons learned and next steps. In *Reflections from the Exploratorium, October 11–12, 2001.* Retrieved January 26, 2005, from www.exploratorium.edu/guidebook/forum/report/.

Ellenbogen, K. M. (2002). Museums in family life: An ethnographic case study. In G. Leinhardt, K. Crowley, & K. Knutson (Eds.), *Learning conversations: Explanation and identity in museums* (pp. 81–101). Mahwah, NJ: Lawrence Erlbaum Associates.

Ellenbogen, K. M. (2003). *From dioramas to the dinner table: An ethnographic case study of the role of science museums in family life.* Dissertation Abstracts International, 64(03), 846A. (University Microfilms No. AAT30–85758)

Ellenbogen, K. M., Kessler, C., & Gillmartin, J. (2003). *Biodiversity 911: Saving life on earth, A World Wildlife Fund traveling exhibition.* Summative Evaluation Report. Annapolis, MD: Institute for Learning Innovation.

Ellenbogen, K. M., Luke, J., & Dierking, L. D. (2004). Family learning research in museums: An emerging disciplinary matrix? In L. D. Dierking, K. M. Ellenbogen, & J. H. Falk (Eds.), *In principle, in practice: Perspectives on a decade of museum learning research (1994–2004). Science Education, 88*, Suppl. 1, 48–58.

Elmesky, R., & Tobin, K. (2005). Expanding our understandings of urban science education by expanding the roles of students as researchers. *Journal of Research in Science Teaching, 42*(7), 807–828.

Emery, A. R. (2002). The integrated museum: A meaningful role in society. *Curator*, *44*(1), 69–82.

Ericsson, K. A., & Simon, H. A. (1993). *Protocol analysis: Verbal reports as data.* Cambridge, MA: MIT Press.

Evans, E. M. (2001). Cognitive and contextual factors in the emergence of diverse belief systems: Creation versus evolution. *Cognitive Psychology*, *42*, 217–266.

Evans, G. (1995). Learning and the physical environment. In J. H. Falk & L. D. Dierking (Eds.), *Public institutions for personal learning* (pp. 119–126). Washington, DC: American Association of Museums.

Falk, J. H. (1988). Museum recollections. In S. Bitgood et al. (Eds.), *Proceedings of first annual visitor studies meeting* (pp. 60–65). Jacksonville, AL: Jacksonville State University.

Falk, J. H. (1993). *Leisure decisions influencing African American use of museums.* Washington, DC: American Association of Museums.

Falk, J. H. (1997). Testing a museum exhibition design assumption: Effect of explicit labeling of exhibit clusters on visitor concept development. *Science Education*, *81*(6), 679–688.

Falk, J. H. (2001a). *Free-choice science education: How we learn science outside of school.* New York: Teachers College Press.

Falk, J. H. (2001b). Free-choice science learning: Framing the discussion. In J. H. Falk (Ed.), *Free-choice science education. How we learn science outside of school* (pp. 3–20). New York: Teachers College Press.

Falk, J. H. (2006). The impact of visit motivation on learning: Using identity as a construct to understand the visitor experience. *Curator*, *49*(2), 151–166.

Falk, J. H., & Adelman, L. M. (2003). Investigating the impact of prior knowledge, experience and interest on aquarium visitor learning. *Journal of Research in Science Teaching*, *40*(2), 163–176.

Falk, J. H., Balling, J. D., & Liversidge, J. (1985). *Information and agenda: Strategies for enhancing the educational value of family visits to a zoological park.* Interim Report, Scholarly Studies No. 1231S4–01. Washington, DC: Smithsonian.

Falk, J. H, Brooks, P., & Amin, R. (2001). Investigating the long-term impact of a science center on its community: The California Science Center L.A.S.E.R. Project. In J. H. Falk (Ed.), *Free-choice science education. How we learn science outside of school* (pp. 115–132). New York: Teachers College Press.

Falk, J. H., & Dierking, L. D. (1990). The relation between visitation frequency and long-term recollection. In S. Bitgood, A. Benefield, & D. Patterson (Eds.), *Visitor studies: Theory, research and practice.* Proceedings of 1990 annual Visitor Studies Conference, Jacksonville, AL, Center for Social Design.

Falk, J. H., & Dierking, L. D. (1992). *The museum experience.* Washington, DC: Whalesback Books.

Falk, J. H., & Dierking, L. D. (Eds.). (1995). *Public institutions for personal learning: Establishing a research agenda.* Washington, DC: American Association of Museums.

Falk, J. H., & Dierking, L. D. (1997). School field trips: Assessing their long term impact. *Curator, 40*(3), 211–218.

Falk, J. H., & Dierking, L. D. (2000). *Learning from museums: Visitor experiences and the making of meaning.* Walnut Creek, CA: AltaMira Press.

Falk, J. H., & Dierking, L. D. (2002). *Lessons without limit: How free-choice learning is transforming education.* Walnut Creek, CA: AltaMira Press.

Falk, J. H., Dierking, L. D., Rennie, L. J., Scott, C., & Cohen Jones, M. (2006). *Investigating the long-term impact of interactives on visitor learning: An exploratory study.* Paper presented at the 2006 National Association for Research in Science Teaching (NARST) conference, San Francisco, CA.

Falk, J. H., Koran, J. J., Dierking, L. D., & Dreblow, L. (1985). Predicting visitor behavior. *Curator*, 28, 249–257.

Falk, J. H., Luke, J., & Abrams, C. (1996). *Women's health: Formative evaluation at the Maryland Science Center.* Unpublished evaluation report. Annapolis, MD: Institute for Learning Innovation.

Falk, J. H., Moussouri, T., & Coulson, D. (1998). The effect of visitors' agendas on museum learning. *Curator, 41*(2), 106–120.

Falk, J. H., Scott, C., Dierking, L. D., Rennie, L. J., & Cohen Jones, M. (2004). Interactives and visitor learning. *Curator, 47*(2), 171–198.

Falk, J. H., & Shepherd, B. K. (2006). *Thriving in the knowledge age: New business models for museums and other cultural organizations.* Walnut Creek, CA: AltaMira Press.

Falk, J. H., & Storksdieck, M. (2005). Using the Contextual Model of Learning to understand visitor learning from a science center exhibition. *Science Education, 89*(5), 744–778.

Falk, J. H., Storksdieck, M., & Dierking, L. D. (in press). An investigation into the nature of public science understanding. *Public Understanding of Science.*

Farmelo, G., & Carding, J. (Eds.). (1997). *Here and now: Contemporary science and technology in museums and science centers.* London: Science Museum.

Fienberg, J., & Leinhardt, G. (2002). Looking through the glass: Reflections of identity in conversations at a history museum. In G. Leinhardt, K. Crowley, & K. Knutson (Eds.), *Learning conversations in museums* (pp. 167–212). Mahwah, NJ: Lawrence Erlbaum Associates.

Fivush, R. (1983). Children's long-term memory for a novel event: An exploratory study. *Merrill-Palmer Quarterly, 30*(3), 303–316.

Franklin, U. (1990). Reflections on science and the citizen. In C. Mungall & D. McLaren (Eds.), *Planet under stress: The challenge of global change* (pp. 267–268). Toronto: Oxford University Press.

Freeman, M. (1993). *Rewriting the self: History, memory, narrative.* London: Routledge.

Galani, A., & Chalmers, M. (2002). *Can you see me? Exploring co-visiting between physical and virtual visitors.* Paper presented at the Museums and the Web annual conference 2002. Retrieved January 20, 2005, from *Papers: Museums and the Web 2002*, www.archimuse.com/mw2002/papers/galani.html.

Gallas, K. (1995). *Talking their way into science: Hearing children's questions and theories, responding with curricula.* New York: Teachers College Press.

Gammon, B. (1999). Everything we currently know about making visitor-friendly mechanical interactive exhibits. *Informal Learning Review, 39*, 1–13.

Gammon, B. (2001). *Assessing learning in museum environments: A practical guide for museum evaluators.* London: Science Museum.

Gardner, H. (1991). *The unschooled mind: How children learn, how schools should teach.* New York: Basic Books.

Gardner, H. (1993a). *Frames of mind: The theory of multiple intelligences* (Tenth anniversary edition). New York: Basic Books.

Gardner, H. (1993b). *Multiple intelligences: The theory in practice.* New York: Basic Books.

Garibay, C. (2004). *Animal Secrets: Bilingual labels formative evaluation.* Unpublished raw data. Portland: Oregon Museum of Science and Industry.

Garibay, C. (2005, July). *Visitor studies and underrepresented audiences: Cultural responsiveness in evaluation.* Paper presented at the 2005 Visitor Studies Conference, Philadelphia, PA.

Garibay, C. (2006a). *Primero la Ciencia remedial evaluation.* January 2006. Unpublished manuscript. Chicago: Chicago Botanic Garden.

Garibay, C. (2006b, Spring). *Latino audience research for the program in Latino history and culture.* Washington, DC: National Museum of American History, Smithsonian Institution.

Garibay, C., Gillmartin, J., & Schaefer, J. (2002). *Park Voyagers summative evaluation.* Unpublished manuscript. Chicago: Museums in the Park.

Gay, L. R., & Airasian, P. (2006). *Educational research: Competencies for analysis and applications* (8th ed.). Upper Saddle River, NJ: Merrill Prentice Hall.

Geertz, C. (1983). *Local knowledge: Further essays in interpretative anthropology.* New York: Basic Books.

Gibson, J. J. (1977). The theory of affordances. In R. E. Shaw & J. Bransford (Eds.), *Perceiving, acting, and knowing.* Hillsdale, NJ: Lawrence Erlbaum Associates.

Gilbert, J., & Priest, M. (1997). Models and discourse: A primary school science class visit to a museum. *Science Education, 81*(6), 749–762.

Gilbert, J., & Priest, M. (2001). What do primary students gain from discussion about exhibits? In S. Errington, S. Stocklmayer, & B. Honeyman (Eds.), *Using museums to popularize science and technology*. Pall Mall, London: Commonwealth Secretariat.

Gill, C. J. (1999). Invisible ubiquity: The surprising relevance of disability issues in evaluation. *American Journal of Evaluation, 20*(2), 279–289.

Gillon, S. (2004). *Boomer nation: The largest and richest generation ever and how it changed America.* New York: Free Press / Simon & Schuster.

Gilmore, J. H., & Pine, B. J., II. (1997). The four faces of mass customization. *Harvard Business Review, 75*(1), 91–101.

Gilmore, J. H., & Pine, B. J., II. (2002). Customer experience places: The new offering frontier. *Strategy & Leadership, 30*(4), 4–11.

Gladwell, M. (2000). *The tipping point: How little things can make a big difference.* Boston: Little, Brown.

Glaser, B., & Strauss, A. (1967). *The discovery of grounded theory: Strategies for qualitative research.* Chicago: Aldine.

Goldsmith, R. E., & Freiden, J. B. (2004). Have it your way: Consumer attitudes toward personalized marketing. *Marketing Intelligence & Planning, 22*(2), 228–239.

Gonzalez, N., & Moll, C. (2002). Cruzando el puente: Building bridges to funds of knowledge. *Educational Policy, 16*(4), 623–41.

Gottfried, J. (1979). *A naturalistic study of children's behavior in a free-choice learning environment.* Unpublished PhD dissertation, University of California, Berkeley.

Green, J., & Dixon, C. (1993). Santa Barbara classroom discourse group. *Linguistics and Education, 5*(3–4), 231–240.

Green, J., & Meyer, L. (1991). The embeddedness of reading in classroom life. In C. Baker & A. Luke (Eds.), *Towards a critical sociology of reading pedagogy* (pp. 141–160). Philadelphia: John Benjamin.

Greene, J. C., Caracelli, V. J., & Graham, W. F. (1989). Toward a conceptual framework for mixed-method evaluation designs. *Educational Evaluation and Policy Analysis, 11*, 255–274.

Griffin, D. J. G., & Abraham, M. (2001). The effective management of museums: Cohesive leadership and visitor-focused public programming. *International Journal of Museum Management and Curatorship. 18*(4), 335–368.

Griffin, D. J. G., Abraham, M., & Crawford, J. (1999). Effective management of museums in the 1990s. *Curator*, *42*(1), 37–53.

Griffin, J. (1994). Learning to learn in informal settings. *Research in Science Education*, *24*, 121–128.

Griffin, J. (1998). *School-museum integrated learning experiences in science: A learning journey.* Unpublished PhD dissertation, University of Technology, Sydney, Australia.

Griffin, J. (1999). Finding evidence of learning in museum settings. In E. Scanlon, E. Whitelegg, & S. Yates (Eds.), *Communicating science: Contexts and channels* (pp. 110–119). London: Routledge.

Griffin, J. (2004). Research on students and museums: Looking more closely at the students in school groups. *Science Education*, *88*(suppl 1), S59–70.

Griffin, J., Meehan, C., & Jay, D. (2003, May). *The other side of evaluating student learning in museums.* Paper presented at the Museums Australia Conference, Perth.

Griffin, J., & Symington, D. (1997). Moving from task-oriented to learning-oriented strategies on school excursions to museums. *Science Education*, *81*(6), 763–779.

Groundwater-Smith, S., & Kelly, L. (2003). As we see it: Improving learning in the museum. *British Educational Research Annual Conference* (pp. 1–22). Edinburgh, Scotland.

Guba, E. G., & Lincoln, Y. S. (1989). *Fourth generation evaluation.* Newbury Park, CA: Sage.

Gumperz, J. (1986). Interactive sociolinguistics in the study of schooling. In J. Cook & J. Gumperz (Eds.), *The social construction of literacy* (pp. 45–68). London: Cambridge.

Gurian, E. H. (1995). *Institutional trauma: Major change in museums and its effect on staff.* Washington, DC: American Association of Museums.

Gurian, E. H. (2004). Singing and dancing at night. In L. E. Sullivan & A. Edwards (Eds.), *Stewards of the sacred* (pp. 89–96). Washington, DC: American Association of Museums / Center for the Study of World Religions, Harvard University.

Gutwill, J. P. (2006). Labels for open-ended exhibits: Using questions and suggestions to motivate physical activity. *Visitor Studies Today*, *9*(1), 1–9.

Gutwill-Wise, J., Soler, C., Allen, S., Wong, D., & Rezny, S. (2000). *Revealing bodies summative evaluation.* Retrieved from www.exploratorium.edu/partner/evaluation.html.

Guy, M., & Kelley-Lowe, M. B. (2001, March). *Preservice elementary teachers' perceptions of a teaching partnership with a museum.* Paper presented at the annual meeting of the National Association for Research in Science Teaching, St. Louis, MO.

Guzzo, R., & Dickerson, M. (1996). Teams in organizations: Recent research on performance and effectiveness. *Annual Review of Psychology*, *47*, 307–338.

Gyllenhaal, E. D. (1998). *Trustworthiness of naturalistic inquiry.* Unpublished manuscript. Chicago: Selinda Research Associates.

Gyllenhaal, E. D., Perry, D. L., & Cheng, B. (2005). *Sugar from the Sun formative evaluation.* Unpublished manuscript. Chicago: Garfield Park Conservatory Alliance.

Hamel, G., & Prahalad, C. K. (1994). *Competing for the future.* Cambridge, MA: HBR Press.

Hart, C. W. L. (1995). Mass customization: Conceptual underpinnings, opportunities and limits. *International Journal of Service Industry Management*, *6*(2), 36–45.

Hayward, D. G., & Brydon-Miller, M. L. (1984). Spatial and conceptual aspects of orientation: Visitor experiences at an outdoor history museum. *Journal of Environmental Systems*, *13*(4), 317–332.

Hedge, A. (1995). Human-factor considerations in the design of museums to optimize their impact on learning. In J. H. Falk & L. D. Dierking (Eds.), *Public institutions for personal learning: Establishing a research agenda* (pp. 105–118). Washington, DC: American Association of Museums.

Heifetz, R., & Linsky, M. (2002). A survival guide for leaders. *Harvard Business Review*, *80*(6), 65–74.

Heimlich, J. E., Bronnenkant, K., Barlage, J., & Falk, J. H. (2005). *Measuring the learning outcomes of adult visitors to zoos and aquariums: Phase I study.* Unpublished technical report. Bethesda, MD: American Association of Zoos and Aquariums.

Heimlich, J. E., Storksdieck, M., Barlage, J., & Falk, J. H. (2005). *Catalina Island Conservancy: A triangulated study of conservation stakeholders.* Annapolis, MD: Institute for Learning Innovation.

Hein, G. E. (1998). *Learning in the museum.* London: Routledge.

Hein, G. E. (2003). *Final summary evaluation report for Traits of Life, a collection of life science exhibits.* Retrieved from www.exploratorium.edu/partner/evaluation.html.

Hein, G. E. (2004). John Dewey and museum education. *Curator*, *47*(4), 413–427.

Heintz, J., Wicks-Lim, J., & Pollin, R. (2005). *Decent work in America.* University of Massachusetts, Amherst. Retrieved November 7, 2005, from www.umass.edu/peri/pdfs/DWA.pdf.

Hemp, P., & Stewart, T. (2004). Leading change when business is good. *Harvard Business Review*, *82*(12), 61–70.

Hilke, D. D. (1987). Museums as resources for family learning: Turning the question around. *Museologist*, *50*(175), 14–15.

Holland, D., Lachicotte, W., Skinner, D., & Cain, C. (1998). *Identity and agency in cultural worlds.* Cambridge, MA: Harvard University Press.

Hood, M. G. (1983). Staying away: Why people choose not to visit museums. *Museum News*, *61*(4), 50–57.

Hooper-Greenhill, E. (2000). *Museums and the interpretation of visual culture.* London: Routledge.

Hooper-Greenhill, E., & Moussouri, T. (2002). *Researching learning in museums and galleries 1990–1999: A bibliographic review.* Leicester, UK: University of Leicester.

Hout, T. M., & Carter, J. C. (1995, November–December). Getting it done: New roles for senior executives. *Harvard Business Review*, 133–146.

Hsi, S. (2002). *Evaluation of electronic guidebook mobile web resources.* Retrieved January 25, 2005, from www.exploratorium.edu/guidebook/eguide_exec_summary_02.pdf.

Hudson, J. W. (1983). Scripts, episodes and autobiographical memories. In K. Nelson (Chair), *Memory and the real world: Symposium conducted at the meetings of the Society for Research in Child Development*, Detroit, MI.

Hui-Min, C., & Cooper, M. D. (2001). Using clustering techniques to detect usage patterns in a web-based information system. *Journal of the American Society for Information Science and Technology*, *52*(11), 888–904.

Humphrey, T., Gutwill, J. P., & the Exploratorium *APE* Team. (2005). *Fostering active prolonged engagement: The art of creating APE exhibits.* San Francisco: Exploratorium.

Hutchins, E. (1995). *Cognition in the wild.* Cambridge, MA: MIT Press.

Hutchins, R. M. (1970). *The learning society.* Harmondsworth: Penguin.

Istomina, Z. M. (1975). The development of voluntary memory in preschool age children. *Soviet Psychology*, *13*, 6–64.

Ivanova, E. (2003). Changes in collective memory: The schematic narrative template of victimhood in Kharkiv museums. *Journal of Museum Education, 28*(1), 17–22.

Iwasyk, M. (2000). Kids questioning kids: "Experts" sharing. In J. Minstrell & E. van Zee (Eds.), *Inquiring into inquiry learning and teaching in science* (pp. 130–138). Washington, DC: American Association for the Advancement of Science.

Janes, R. R. (1997). *Museums and the paradox of change: A case study in urgent adaptation* (2nd ed.). Calgary, AB: Glenbow Museum.

Janousek, I. (2000). The "context museum": Integrating science and culture. *Museum International*, *52*(4), 21–24.

Jeffery-Clay, K. R. (1998). Constructivism in museums: How museums create meaningful learning environments. *Journal of Museum Education*, *23*(1), 3–7.

Jennings, G. (Chair). (2005, October). *Science as a world view: Can science explain everything?* Panel presentation at the Association of Science and Technology Centers conference, Richmond, VA.

Jensen, N. (1994). Children's perceptions of their museum experiences: A contextual perspective. *Children's Environments, 11*, 300–324.

Jensen, N. (1995, May). *Children's perceptions of their museum experiences: A contextual perspective.* Paper presented at *Current Trends in Audience Research and Evaluation*, Philadelphia, PA.

Johnson, R. B., & Onwuegbuzie, A. J. (2004). Mixed method research: A research paradigm whose time has come. *Educational Researcher, 33*(7), 14–26.

Johnston, D. (1999). *Assessing the visiting public's perceptions of the outcomes of their visit to interactive science and technology centres.* Unpublished PhD thesis, Curtin University of Technology, Perth, Western Australia.

Johnston, D., & Rennie, L. J. (1995). Perceptions of visitors' learning in an interactive science and technology centre in Australia. *Museum Management and Curatorship, 14*, 317–325.

Johnston, J. (2005, March 22). Ministry uses dinosaurs to dispute evolution. *Enquirer.* Retrieved June 7, 2005, from http://news.enquirer.com.

Jones, D., & Stein, J. K. (2005). *The Flandrau Science Center front-end evaluation.* Unpublished technical report. Annapolis, MD: Institute for Learning Innovation.

Karp, I. (1991). Cultures in museum perspective. In I. Karp and S. Levine (Eds.), *Exhibiting cultures: The poetics and politics of museum display* (pp. 373–385). Washington, DC: Smithsonian Institution Press.

Karp, I., & Levine, S. (Eds.). (1991). *Exhibiting cultures: The poetics and politics of museum display.* Washington, DC: Smithsonian Institution Press.

Kirkhart, K. E. (1995). Seeking multicultural validity: A postcard from the road. *Evaluation Practice, 16*, 1–12.

Kisiel, J. (2001, March 27). *Worksheets, museums and teacher agendas: A closer look at a learning experience.* Paper presented at the annual meeting of National Association for Research in Science Teaching, St. Louis, MO.

Klages, E., & the Exploratorium Staff. (1999). *Facilitating the framework.* San Francisco: Exploratorium.

Kofman, F., & Senge, P. (1995). Communities of commitment: The heart of learning organizations. In S. Chawla & J. Renesch (Eds.), *Learning organizations: Developing cultures for tomorrow's workplace* (pp. 24–28). Portland, OR: Productivity Press.

Kolb, D. A. (1984). *Experiential learning: Experience as a source of learning and development.* New Jersey: Prentice Hall.

Korn, R. (2005, October). The intentional museum. In G. Jennings (Chair), *Science as a world view: Can science explain everything?* Panel presentation at the Association of Science and Technology Centers conference, Richmond, VA.

Koster, E. H. (1999). In search of relevance: Science centers as innovators in the evolution of museums. *Daedalus*, *128*(3), 277–296.

Koster, E. H. (2006). The Relevant Museum: A Reflection on Sustainability. *Museum News*, May/June 2006, 67–90.

Koster, E. H. (2002) A Tragedy Revisited. *Muse: Journal of the Canadian Museum Association*, *20*, 26–27.

Koster, E. H., & Baumann, S. H. (2005). Liberty Science Center in the United States: A mission focused on external relevance. In R. Janes and G. Conaty (Eds.), *Looking reality in the eye: Museums and social responsibility* (pp. 85–111). Calgary, AB: University of Calgary.

Kress, G. (1995). *Making signs and making subjects: The English curriculum and social futures.* London: Institute of Education.

Kubota, C. A., & Olstad, R. G. (1991). Effects of novelty-reducing preparation on exploratory behavior and cognitive learning in a science museum setting. *Journal of Research in Science Teaching*, *28*(3), 225–234.

Kuhn, T. S. (1970). *The structure of scientific revolutions.* Chicago: University of Chicago Press.

Laboratory of Comparative Human Cognition. (1983). Culture and cognition development. In W. Kessen (Ed.), *Manual of child psychology: History, theories and methods* (Vol. 1). New York: Wiley.

Latane, B., & Darley, J. (1969). Bystander "apathy." *American Scientist*, *57*, 244–268.

Lave, J. (1993). The practice of learning. In J. Lave & S. Chaiklin (Eds.), *Understanding practice: Perspectives on activity and context* (pp. 3–32). New York: Cambridge University Press.

Lave, J., & Wenger, E. (1991). *Situated learning: Legitimate peripheral participation.* Cambridge: Cambridge University Press.

Lebeau, R., Gyamfi, P., Wizevich, K., & Koster, E. (2001). Supporting and documenting choice in free-choice science learning environments. In J. H. Falk (Ed.), *Free-choice science education: How we learn science outside of school* (pp. 133–148). New York: Teachers College Press.

Lee, D. (1959). *Freedom and culture.* New York: Prentice Hall.

Leinhardt, G., Crowley, K., & Knutson, K. (Eds.). (2002). *Learning conversations in museums.* Mahwah, NJ: Lawrence Erlbaum Associates.

Leinhardt, G., & Knutson, K. (2004). *Listening in on museum conversations.* Walnut Creek, CA: AltaMira Press.

Leinhardt, G., Knutson, K., & Crowley, K. (2003). Museum learning collaborative redux. *Journal of Museum Education*, *28*(1), 23–31.

Leinhardt, G., Tittle, C., & Knutson, K. (2002). Talking to oneself: Diaries of museum visits. In G. Leinhardt, K. Crowley, & K. Knutson (Eds.), *Learning conversations in museums* (pp. 103–133). Mahwah, NJ: Lawrence Erlbaum Associates.

Lemke, J. L. (1993). *Talking science: Language, learning and values.* Norwood, NJ: Ablex.

Lemonick, M. D. (2006). Is America flunking science? *Time*, *167*(7), 22–38.

Levy, B. H. (2006, February 3). Lecture at Politics and Prose Bookstore, Washington, DC.

Lewenstein, B. V. (2001). Who produces scientific information for the public? In J. H. Falk (Ed.), *Free-choice science education: How we learn science outside of school* (pp. 21–43). New York: Teachers College Press.

Lewis, M. W., & Kelemen, M. L. (2002). Multiparadigm inquiry. *Human Relations*, *55*(2), 251–275.

Lincoln, Y. S., & Guba, E. G. (1985). *Naturalistic inquiry.* Newbury Park, CA: Sage.

Livingstone, P., Pedretti, E., & Soren, B. (2001). Visitor comments and the sociocultural context of science: Public perceptions and the exhibition A Question of Truth. *Museum Management and Curatorship*, *19*(4), 355–369.

Louv, R. (2005). *Last child in the woods: Saving our children from nature deficit disorder.* Chapel Hill, NC: Algonquin Books.

Lublin, J. S. (1993, July 20). Best manufacturers found to triumph by fostering co-operation among employers. *Wall Street Journal*, A2.

Lucas, A. M., McManus, P., & Thomas, G. (1986). Investigating learning from informal sources: Listening to conversations and observing play in science museums. *European Journal of Science Education*, *8*, 341–352.

Luke, J. J., Bronnenkant, K., & Dierking, L. (2003). *Parent Partners in School Science: Year 2 final report.* Unpublished evaluation report. Annapolis, MD: Institute for Learning Innovation.

Luke, J. J., Büchner, K., Dierking, L. D., & O'Ryan, B. (1999). *Creative World summative evaluation: California Science Center.* Unpublished evaluation report. Annapolis, MD: Institute for Learning Innovation.

Luke, J. J., Cohen Jones, M., Dierking, L., Adams, M., & Falk, J. (2002). *The Children's Museum of Indianapolis Family Learning Initiative: Phase II programs study.* Unpublished technical report. Annapolis, MD: Institute for Learning Innovation.

Luke, J. J., Coles, U., & Falk, J. (1998). *Summative evaluation of DNA Zone, St. Louis Science Center.* Technical report. Annapolis, MD: Institute for Learning Innovation.

Luke, J. J., O'Mara, H., & Dierking, L. (1999). *Mammals Hall front-end evaluation Phase II, National Museum of Natural History.* Unpublished evaluation report. Annapolis, MD: Institute for Learning Innovation.

Macdonald, S. (1998). *The politics of display: Museums, science, culture.* New York: Routledge.

Machlup, F. (1962). *The production and distribution of knowledge in the U.S.* Princeton, NJ: Princeton University Press.

Mallwitz, S. (2006). The future of the museum? *The Informal Learning Review, 75*(November–December), 20–22.

Manchester Guardian. (2005, September). Letter to editor.

Marshall, C. (2002). *Museums as catalysts for interdisciplinary collaboration.* Cambridge, MA: Museum Loan Network at Massachusetts Institute for Technology.

Martin, L. (2001). Free-choice science learning: Future directions for researchers. In J. H. Falk (Ed.), *Free-choice science education: How we learn science out of school.* New York: Teachers College Press.

Martin, L. M. W., & Toon, R. (2005). Narratives in a science center: Interpretations and identity. *Curator, 48*(4), 407–425.

Mathewson, D., & McKeon, P. (2002, December). *Disrupting notions of collaboration: The problematic engagement of museums and schools.* Paper presented at the Australian Association for Research in Education, Brisbane, Australia.

Mathison, S. (1988). Why triangulate? *Educational Researcher, 17*(2), 13–17.

Matusov, E., Bell, N., & Rogoff, B. (2002). Schooling as cultural process: Shared thinking and guidance by children from schools differing in collaborative practices. In R. Kail & H. Reese (Eds.), *Advances in Child Development and Behavior, 29*, 129–160.

Matusov, E., & Rogoff, B. (1995). Evidence of development from people's participation in communities of learners. In J. H. Falk & L. D. Dierking (Eds.), *Public institutions for personal learning: Establishing a research agenda* (pp. 97–104). Washington, DC: American Association of Museums.

Maxwell, L. E., & Evans, G. W. (2002). Museums as learning settings: The importance of the physical environment. *Journal of Museum Education, 27*(1), 3–7.

McCarthy, B. (1987). *The 4MAT System: Teaching to learning styles with right/left mode techniques.* Barrington, IL: EXCEL.

McCombs, B. (1991). Motivation and lifelong learning. *Educational Psychologist, 26*(2), 117–127.

McCombs, B. (1996). Alternative perspectives for motivation. In L. Baker, P. Afflerback, & D. Reinking (Eds.), *Developing engaged readers in school and home communities* (pp. 67–87). Mahwah, NJ: Lawrence Erlbaum Associates.

McConnell, M., & Hess, H. (1998). A controversy timeline. *Journal of Museum Education*, *23*(3), 4–6.

McCrory, P. (2002, Autumn). Blurring the boundaries between science centers and schools. *ECSITE Newsletter*, *52*, 11–12.

McLaughlin, H. (1998). The pursuit of memory: Museums and the denial of the fulfilling sensory experience. *Journal of Museum Education*, *23*(3), 10–12.

McLean, F. (2004). Museums and national identity. *Museum and Society*, *3*(1), 1–4.

McLean, K. (1999). Opening up the Exploratorium. *Journal of Museum Education*, *24*(1/2), 11–15.

McLean, K. (Ed.). (2003). Visitor voices. *Journal of Museum Education*, *28*(3).

McLean, K., & McEver, C. (Eds.). (2004). *Are we there yet? Conversations about best practices in science exhibition development.* San Francisco: Exploratorium.

McLeod, J., & Kilpatrick, K. (2002). *Exploring science at the museum.* Reston, VA: Association for Supervision and Curriculum Development.

McManus, P. (1987). It's the company you keep . . . the social determination of learning-related behavior in a science museum. *International Journal of Museum Management and Curatorship*, *6*, 263–270.

McManus, P. (1988). Good companions: More on the social determination of learning-related behavior in a science museum. *International Journal of Museum Management and Curatorship*, *7*, 37–44.

McManus, P. (1993). Memories as indicators of the impact of museum visits. *Museum Management and Curatorship*, *12*, 367–380.

Medved, M. (1998). *Remembering exhibits at museums of art, science and sport.* Unpublished PhD thesis. University of Toronto.

Medved, M., Cupchik, G., & Oatley, K. (2004). Interpretive memories of artworks. *Memory*, *12*(1), 119–128.

Medved, M., & Oatley, K. (2000). Memories and scientific literacy: Remembering exhibits from a science centre. *International Journal of Science Education*, *22*(10), 1117–1132.

Melton, A. W., Feldman, N. G., & Mason, C. W. (1936). *Experimental studies of the education of children in a museum of science.* Washington, DC: American Association of Museums.

Merriman, N. (1991). *Beyond the glass case.* Leicester, UK: Leicester University Press.

Millett, R. (2002). Missing voices: A personal perspective on diversity in program evaluation. *Non-profit Quarterly E-newsletter*, *1*(12). Retrieved June 18, 2006, from www.nonprofitquarterly.org/section/309.html.

Mintz, A. (1995). *Communicating controversy: Science museums and issues education.* Washington, DC: Association of Science-Technology Centers.

Mok, C., Stutts, A. T., & Wong, L. (2000). *Mass customization in the hospitality industry: Concepts and Applications.* Retrieved January 19, 2005, from www.hotel-online.com/Neo/Trends/ChiangMaiJun00/CustomizationHospitality.html.

Monroe, C., Lau, A., Schubel, J. R., & Cassano, E. (2006). *Public ocean literacy: Making ocean science understandable.* MCRI Aquatic Forum Report Reference Number 2006–2. Long Beach, CA: Marine Conservation Research Institute.

Moore, M. (1995). *Creating public value: Strategic management in government.* Cambridge, MA: Harvard University Press.

Moussouri, T. (1997). *Family agendas and family learning in hands-on science museums.* PhD dissertation, University of Leicester, England.

Muir, D. (2005). *National myth of the American Indian.* Retrieved January 13, 2006, from www.freerepublic.com/focus/f-news/1358787/posts.

Mumford, M. D., & Licuanan, B. (2004). Leading for innovation: Conclusions, issues, and directions. *Leadership Quarterly, 15*, 163–171.

Munley, M. E., & Roberts, R. (2006). Are museum educators still necessary? *Journal of Museum Education, 31*(1), 29–40.

Murray, V., & Tassie, B. (1994). Evaluating the effectiveness of nonprofit organizations. In Robert D. Herman & Associates (Eds.), *The Jossey-Bass Handbook of Nonprofit Leadership and Management* (Vol. 14, pp. 303–324). San Francisco: Jossey-Bass.

Museum Loan Network. (2002). *Museum as catalyst for interdisciplinary collaboration: Beginning a conversation.* Cambridge, MA: Museum Loan Network.

Nader, K. (2003). Memory traces unbound. *TRENDS in Neurosciences, 28*(2), 65–72.

National Academies, The. (2005). *Rising above the gathering storm: Energizing and employing America for a brighter economic future.* National Academy of Sciences, National Academy of Engineering, and Institute of Medicine, December 2005. Retrieved from www.national-academies.org.

National Commission on Mathematics and Science Teaching for the 21st Century. (2000). *Before it's too late.* Washington, DC: U.S. Department of Education. Retrieved from www.ed.gov/americacounts/glenn.

National Museum of the American Indian. Mission statement on website. Retrieved January 21, 2005, from www.nmai.si.edu/subpage.cfm?subpage=visitor&second=about&third=about.

National Research Council. (1999). *How people learn: Brain, mind, experience, and school.* Washington, DC: National Academy Press.

National Research Council. (2002). *Scientific research in education.* Washington, DC: National Academy Press.

National Science Board. (2005). *2020 Vision for the National Science Foundation.* Retrieved December 28, 2005, from www.nsf.gov/publications/pub_summ.jsp?ods_key=nsb05142.

Neisser, U., & Fivush, R. (1994). *The remembering self: Construction and accuracy in self-narratives.* New York: Cambridge University Press.

Newlands, D. L. (1983, Summer). Stress and distress in museum work. *Muse* (1–2), 18–33.

Norman, D. A. (1988). *The design of everyday things.* New York: Doubleday.

Office of Policy and Analysis. (2002). *21st century roles of national museums: A conversation in progress.* Washington, DC: Smithsonian Institution.

Ogbu, J. U. (1995). The influence of culture on learning and behavior. In J. H. Falk & L. D. Dierking (Eds.), *Public institutions for personal learning* (pp. 79–95). Washington, DC: American Association of Museums.

Ogden, J. L., Lindburg, D. G., & Maple, T. L. (1993). The effects of ecologically relevant sounds on zoo visitors. *Curator, 36*(2), 147–156.

Orion, N. (1993). A model for the development and implementation of field trips as an integral part of the science curriculum. *School Science and Mathematics, 93*(6), 325–331.

Orion, N. (1999, March). *A holistic approach in making outdoor activities an integral part of the formal learning process in schools.* Paper presented at the annual meeting of National Association for Research in Science Teaching, Boston, MA.

Palys, T. (1997). *Research decisions: Quantitative and qualitative perspectives.* (2nd ed.). Toronto, ON: Harcourt Canada.

Papanek, V. (1971). *Design for the real world: Human ecology and social change.* New York: Pantheon.

Paris, S. (1997). Situated motivation and informal learning. *Journal of Museum Education, 22*(2&3), 22–26.

Paris, S. G. (Ed.). (2002). *Perspectives on object-centered learning in museums.* Mahwah, NJ: Lawrence Erlbaum Associates.

Paris, S. G., & Ash, D. (2000). Reciprocal theory building inside and outside museums. *Curator, 43*(3), 199–210.

Paris, S. G., & Mercer, M. (2002). Finding self in objects: Identity exploration in museums. In G. Leinhard, K. Crowley, & K. Knutson (Eds.), *Learning conversations in museums* (pp. 401–423). Mahwah, NJ: Lawrence Erlbaum Associates.

Paris, S., Yambor, K., & Wai-Ling Packard, B. (1998). Hands-on biology: A museum-school-university partnership for enhancing students' interest and learning in science. *Elementary School Journal, 98*(3), 267–289.

Parsons, C., & Muhs, K. (1994). Field trips and parent chaperones: A study of self-guided school groups at the Monterey Bay Aquarium. *Visitor Studies: Theory, Research and Practice, 7*(1), 57–61.

Patton, M. Q. (2002). *Qualitative research and evaluation methods.* Thousand Oaks, CA: Sage.

Pearce, M. (2003). "They said that the glass is full of friendship": Visitor stories in a memory exhibition. *Journal of Museum Education*, *28*(3), 26–30.

Pearce, S. (Ed.). (1994). *Interpreting objects and collections.* London: Routledge.

Pedretti, E. (1999). Decision-making and STS education: Exploring scientific knowledge and social responsibility in schools and science centers through an issues-based approach. *School Science and Mathematics*, *99*(4), 174–181.

Pedretti, E. (2002). T. Kuhn meets T. Rex: Critical conversations and new directions in science centers and science museums. *Studies in Science Education*, *37*, 1–42.

Pedretti, E., & Forbes, J. (2000). A question of truth: Critiquing the culture and practice of science through science centers and schools. In D. Hodson (Ed.), *OISE papers in STSE education* (Vol. 1, pp. 87–106). Toronto, ON: University of Toronto Press.

Pedretti, E., & Hodson, D. (1995). From rhetoric to action: Implementing STSE education through action research. *Journal of Research in Science Teaching*, *32*(5), 463–486.

Pedretti, E., Macdonald, R., Gitari, W., & McLaughlin, H. (2001). Visitor perspectives on the nature and practice of science: Challenging beliefs through A Question of Truth. *Canadian Journal of Science, Mathematics and Technology Education*, *1*(4), 399–418.

Pedretti, E., & Soren, B. (2003). A Question of Truth: A cacophony of visitor voices. *Journal of Museum Education*, *28*(3), 17–20.

Pekarik, A. J., Doering, Z. D., & Karns, D. A. (1999). Exploring satisfying experiences in museums. *Curator*, *42*(2), 152–173.

Perry, D. (1989). The creation and verification of a development model for the design of a museum exhibit. *Dissertation Abstracts International*, *50*(12A), p. 3926 (UMI No. 9012186).

Perry, D. (1993). Measuring learning with the knowledge hierarchy. *Visitor Studies: Theory, Research and Practice*, *6*, 73–77.

Perry, D., & Morrissey, K. (2005, May 20). *Designing for conversation.* Workshop presented at the Garfield Park Conservatory, Chicago.

Pew Oceans Commission. (2003). *America's living ocean: Charting a course for sea change.* Arlington, VA: Author.

Pfeffer, J. (1996). When it comes to "best practices"—Why do smart organizations occasionally do dumb things? *Organizational Dynamics*, *25*(1), 33–44.

Phillips, W., & Case, M. (2005). Museums grow old and calcify: Here's why. *QM2 Management Briefing.* Retrieved November 11, 2005, from www.qm2.0rg/mbriefs/38.html.

Piaget, J. (1952). *The origins of intelligence in children.* (M. Cook, Trans.). New York: International Universities Press.

Pine, B. J., II. (1993). *Mass customization: The new frontier in business competition.* Boston: Harvard Business School Press.

Pine, B. J., II. (1998). You're only as agile as your customers think. *Agility & Global Competition*, *2*(2), 24–35.

Pine, B. J., II. (2002). *The Power of Intentional Design.* Retrieved January 21, 2005, from www.DesignIntelligence.com.

Pine, B. J., II, Peppers, D., & Rogers, M. (1995, March–April). Do you want to keep your customers forever? *Harvard Business Review*, 103–114.

Pott, P. (1963). The role of museums of history and folklore in a changing world. *Curator*, *6*(2), 157–170.

Prentice, R., Davies, A., & Beeho, A. (1997). Seeking generic motivations for visiting and not visiting museums and like cultural attractions. *Museum Management and Curatorship*, *6*, 45–70.

Prosavac, E. J. (1998). Toward more informative use of statistics: Alternatives for program evaluators. *Evaluation and Program Planning*, *21*, 243–254.

Rand, J. (2001). Forum: The 227-mile museum, or a visitors' bill of rights. *Curator*, *44*(1), 7–14.

Randol, S. (2004). *Looking for inquiry: Developing an instrument to assess inquiry at museum exhibits.* Proceedings from annual meeting of American Educational Research Association (pp. 1–22). San Diego, California.

Randol, S. M. (2005). *The nature of inquiry in science centers: Describing and assessing inquiry at exhibits.* Unpublished PhD dissertation, University of California, Berkeley.

Rayport, J. F. (2005, February). Demand-side innovation. *Harvard Business Review*.

Reich, C. A. (2000, July–August). The power of universal design: Building an accessible exhibition. *Dimensions*, 14–15.

Reich, C. A. (2005). *Universal design of interactives for museum exhibitions.* Unpublished master's thesis, Lesley University, Cambridge, MA.

Rennie, L. J., Feher, E., Dierking, L. D., & Falk, J. H. (2003). Toward an agenda for advancing research on science learning in out-of-school settings. *Journal of Research in Science Teaching*, *40*, 112–120.

Rennie, L. J., & Johnston, D. (1997). What can floor staff tell us about visitor learning? *Museum National*, *5*(4), 17–18.

Rennie, L. J., & McClafferty, T. P. (2002). Objects and learning: Understanding young children's interaction with science exhibits. In S. Paris (Ed.), *Perspectives on object-centered learning in museums* (pp. 191–214). Mahwah, NJ: Lawrence Erlbaum Associates.

Rennie, L. J., & Stocklmayer, S. M. (2003). The communication of science and technology: Past, present and future agendas. *International Journal of Science Education*, *25*, 759–773.

Rennie, L. J., & Williams, G. F. (2002). Science centers and scientific literacy: Promoting a relationship with science. *Science Education*, *86*, 706–726.

Resnick, L. B. (1987). The 1987 presidential address: Learning in school and out. *Educational Researcher*, *16*(9), 13–20.

Resnick, L. B., Levine, J. M., & Teasley, S. D. (1991). *Perspectives on socially shared cognition.* Washington, DC: American Psychological Association.

Richards, W. H., & Menninger, M. (1993). A discovery room for adults. *Journal of Museum Education*, *19*(1), 6–11.

Ringaert, L. (2001). User/expert involvement in universal design. In W. F. E. Preiser & E. Ostroff (Eds.), *Universal design handbook* (pp. 6.1–6.14). New York: McGraw-Hill.

Roberts, L. C. (1997). *From knowledge to narrative: Educators and the changing museum.* Washington, DC: Smithsonian Institution Press.

Robins, C., & Woollard, V. (2003). *Creative connections: Working with teachers to use museums and galleries as a learning resource.* London: Victoria and Albert Museum and Institute of Education, University of London.

Robinson, E. S. (1928). *The behavior of the museum visitor*. (New series, Number 5). Washington, DC: American Association of Museums.

Rogoff, B. (1998). Cognition as a collaborative Process. In W. Damon, D. Kuhn, & R. Siegler (Eds.), *Handbook of child psychology* (5th ed., Vol. 2, pp. 679–744). New York: Wiley & Sons.

Rogoff, B. (2003a). *The cultural nature of human development.* New York: Oxford University Press.

Rogoff, B. (2003b). *The Center for Informal Learning and Schools: Current research.* Presentation at the annual meeting of the Association of Science-Technology Centers, St. Paul, MN.

Rogoff, B., & Lave, J. (Eds.). (1984). *Everyday cognition: Its development in social contexts.* Cambridge, MA: Harvard University Press.

Rogoff, B., Paradise, R., Mejia Arauz, R., Correa-Chávez, M., & Andelillo, C. (2003). Firsthand learning through intent participation. *Annual Review of Psychology*, *54*, 175–203.

Roschelle, J. (1995). Learning in interactive environments: Prior knowledge and new experience. In J. H. Falk & L. D. Dierking (Eds.), *Public institutions for personal learning: Establishing a research agenda* (pp. 37–51). Washington, DC: American Association of Museums.

Rosebery, A., Warren, B., & Conant, F. (1989). *Making sense of science in language minority classrooms* (Technical Report No. 7306). Cambridge, MA: Bolt Baranek and Newman.

Rosebery, A., Warren, B., & Conant, F. (1992). Appropriating scientific discourse: Findings from language minority classrooms. *Journal of the Learning Sciences*, *2*, 61–94.

Rosenthal, E., & Blankman-Hetrick, J. (2002). Conversations across time: Family learning in a living history museum. In G. Leinhardt, K. Crowley, & K. Knutson (Eds.), *Learning conversations in museums* (pp. 305–330). Mahwah, NJ: Lawrence Erlbaum Associates.

Rounds, J. (2002). Storytelling in science exhibits. *Exhibitionist, 21*(2), 40–43.

Rounds, J. (2004). Strategies for the curiosity-driven museum visitor. *Curator, 47*(4), 389–412.

Rounds, J. (2006). Doing identity work in museums. *Curator, 49*(2), 133–150.

Rowen, B. & Miskel, C. G. (1999). Institutional theory and the study of educational organizations. In J. Murphy & K. S. Lewis (Eds.), *Handbook of research on educational administration, 2nd edition*. San Francisco: Jossey-Bass, 359–382.

Rowe, S. (2002). The roles of objects in active, distributed, meaning making. In S. Paris (Ed.), *Perspectives on object-centered learning in museums* (pp. 19–36). Mahwah, NJ: Lawrence Erlbaum Associates.

Rudman, P., Sharples, M., & Baber, C. (2002). Supporting learning in using personal technologies. *Proceedings of the European Workshop on Mobile and Contextual Learning*, 44–46.

Sacco, J. C. (1999, May). *Crafting exhibit experiences for school audiences.* Paper presented at the American Association of Museums 94th annual meeting, Cleveland, OH.

Sandifer, C. (2003). Technological novelty and open-endedness: Two characteristics of interactive exhibits that contribute to the holding of visitor attention in a science museum. *Journal of Research in Science Teaching, 40*(2), 121–137.

Sauber, C. M. (Ed.). (1994). *Experiment bench project: A workbook for building experimental physics exhibits.* St. Paul, MN: Science Museum of Minnesota.

Schacter, D. L. (1999). The seven sins of memory: Insights from psychology and cognitive neuroscience. *American Psychologist, 54*, 182–203.

Schacter, D. L. (2001). *The seven sins of memory: How the mind forgets and remembers.* New York: Houghton Mifflin.

Schaefer, J., Perry, D. L., & Gyllenhaal, E. D. (2002). *Underground adventure: Final summative/remedial evaluation.* Unpublished manuscript. Chicago: Field Museum.

Schauble, L., Banks, D. B., Coates, G. D., Martin, L. M. W., & Sterling, P. V. (1996). Outside the classroom walls: Learning in informal environments. In L. Schauble & R. Glaser (Eds.), *Innovations in learning: New environments for education* (pp. 5–24). Mahwah, NJ: Lawrence Erlbaum Associates.

Schauble, L. & Barlett, K. (1997). Constructing a science gallery for children and families: The role of research in an innovative design process. *Science Education, 81*(6), 781–793.

Schauble, L., & Glaser, R. (Eds.). (1996). *Innovations in learning: New environments for education.* Hillsdale, NJ: Lawrence Erlbaum Associates.

Schauble, L., Glaser, R., Raghaven, K., & Reiner, M. (1991). Causal models and experimentation strategies in scientific reasoning. *Journal of the Learning Sciences*, 1(2), 201–238.

Schauble, L., Gleason, M., Lehrer, R., Bartlett, K., Petrosino, A., Allen, A., Clinton, K., Ho, E., Jones, M., Lee, Y., Phillips, J., Siegler, J., & Street, J. (2002). Supporting science learning in museums. In G. Leinhardt, K. Crowley, & K. Knutson (Eds.), *Learning conversations in museums* (pp. 425–452). Mahwah, NJ: Lawrence Erlbaum Associates.

Schauble, L., Leinhardt, G., & Martin, L. (1997). A framework for organizing a cumulative research agenda in informal learning contexts. *Journal of Museum Education*, *22*(2&3), 3–8.

Schneider, B. (executive producer). (1970). *Five easy pieces* [Motion Picture]. Los Angeles: Columbia Pictures, BBS Production.

Schoenfeld, A. L. (1999). Looking toward the 21st century: Challenges of educational theory and practice. *Educational Researcher*, *28*(7), 4–14.

Schubel, J. R., Monroe, C., & Lau, A. (2005). *Public ocean literacy: What residents of Southern California should know*. MCRI Aquatic Forum Report Reference Number 2005–2. Long Beach, CA: Marine Conservation Research Institute.

Science at the Crossroads (editorial). (2005, September). *Crossroads for Planet Earth.* [Special issue]. *Scientific American*, 10.

Scott, E. (2000). *The creation/evolution continuum.* National Center for Science Education. Retrieved November 11, 2005, from www.ncseweb.org.

Scott, E. (2005, October 15). *Lessons from the creationism/evolution controversy.* Presentation at the Association of Science and Technology Centers annual conference, Richmond, VA.

Scribner, S. (1984). Studying working intelligence. In B. Rogoff & J. Lave (Eds.), *Everyday cognition: Its development in social context* (pp. 9–40). Cambridge, MA: Harvard University Press.

Scribner, S. (1985). Thinking in action: Some characteristics of practical thought. In R. J. Sternberg & R. K. Wagner (Eds.), *Practical intelligence.* New York: Cambridge University Press.

Scribner, S. (1990). A sociocultural approach to the study of mind. In G. Greenberg & E. Tobach (Eds.), *Theories of evolution and knowing* (pp. 107–120). Hillsdale, NJ: Lawrence Erlbaum Associates.

Scribner, S., & Cole, M. (1973). Cognitive consequences of formal and informal education. *Science*, *182*(4112), 553–559.

Scribner, S., & Cole, M. (1981). *The psychology of literacy.* Cambridge, MA: Harvard University Press.

Sedzielarz, M. (2003). Watching the chaperones: An ethnographic study of adult-child interactions in school field trips. *Journal of Museum Education*, *28*(2), 20–24.

Seig, M. T. & Blankman-Hetrick, J. (2006, July). Reframing authenticity. Session presented at the meeting of Visitor Studies Association, Grand Rapids, MI.

Seig, M. T., & Bubp, K. (in progress). The culture of empowerment: Enabling front line staff to create and sustain change within an organization.

Serrell, B. (1996). *Exhibit labels: An interpretive approach.* Walnut Creek, CA: AltaMira Press.

Serrell, B. (1997). Paying attention: The duration and allocation of visitors' time in museum exhibitions. *Curator, 40*(2), 108–125.

Serrell, B. (2000). Does cueing visitors increase the time they spend in a museum exhibition? *Visitor Studies Today 3*(2), 3–6.

Shavelson, R., Phillips, D. C., Towne, L., & Feuer, M. J. (2003). On the science of education design studies. *Educational Researcher, 32*(1), 25–28.

Shavelson, R., & Towne, L. (2004). What drives scientific research in education: Questions, not methods, should drive the enterprise. *APS Observer, 17*(4).

Shelnut, S. (2000). Long term museum programs for youth. In J. Hirsch & L. Silverman (Eds.), *Transforming practice.* Washington, DC: Museum Education Roundtable.

Sheppard, B. (2000). *Museums, libraries and the 21st century learner.* Washington, DC: Institute of Museum and Library Services.

Shipman, P. (1994). *The evolution of racism: Human differences and the use and abuse of science.* New York: Simon & Schuster.

Silverman, L. (1990). *"Of us and other things": The content and functions of talk by adult visitor pairs in an art and a history museum.* Unpublished PhD dissertation, University of Pennsylvania, Philadelphia.

Silverman, L. (1995). Visitor meaning-making in museums for a new age. *Curator, 38*(3), 161–170.

Silverman, L., & O'Neill, M. (2004). Change and complexity in the 21st-century museum. *Museum News, 83*(6), 36–43.

Sjöstrand, S-E. (1991). The organization of non-profit activities. SSE/EFI Working paper series in business administration, No. 1999:7. Stockholm: Stockholm School of Economics.

Somerset-Ward, R. (2000). *Connecting communities: Public media in the digital age.* Washington, DC: Benton Foundation.

Soren, B. (1995). Triangulation strategies and images of museums as sites for lifelong learning. *Museum Management and Curatorship, 14*, 31–46.

Spasojevic, M., & Kindberg, T. (2001). *A study of an augmented museum experience.* HPL Technical Report HPL-2001–178.

Spock, M. (1999). The stories we tell about meaning making. *Exhibitionist, 18*(2), 30–34

Spock, M. (2000a). When I grow up I want to work in a place like this. *Curator, 41*(1), 19–30.

Spock, M. (2000b). *A study guide to Philadelphia stories: A collection of pivotal museum memories.* Washington, DC: American Association of Museums.

Spock, M. (in press). Tradeoffs between near- and long-term assessments of learning outcomes. *Journal of Interpretive Research.*

St. John, M., & Perry, D. (1993). A framework for evaluation and research: science, infrastructure and relationships. In S. Bicknell & G. Farmelo (Eds.), *Museums visitor studies in the 90s* (pp. 59–66). London: Science Museum.

St. John, M., & Perry, D. (1996). *An invisible infrastructure: Institutions of informal science education* (Vol. 1). Washington, DC: Association of Science-Technology Centers.

Stein, J., Dierking, L. D., Falk, J. H., & Ellenbogen, K. M. (Eds.). (2006). *Insights: A museum learning resource.* Available from the Institute for Learning Innovation, 166 West Street, Annapolis, MD 21401. Online at www.ilinet.org/ipip/In_Principle_In_Practice_Insights_-_Museum_Learning_Resource.pdf.

Stevens, R., & Hall, R. (1997). Seeing tornado: How videotraces mediate visitor understandings of (natural?) phenomena in a science museum. *Science Education, 81*(6), 735–747.

Stevenson, J. (1991). The long-term impact of interactive exhibits. *International Journal of Science Education, 13*(5), 521–531.

Storksdieck, M. (2001). Differences in teachers' and students' museum field-trip experiences. *Visitor Studies Today, 4*, 8–12.

Storksdieck, M. (2006). *Field trips in environmental education.* Berlin, Germany: Berliner Wissenschafts-Verlag.

Storksdieck, M., Ellenbogen, K. M., & Heimlich, J. E. (2005). Changing minds? Factors that influence free-choice learning about environmental conservation. *Environmental Education Research, 11*(3), 353–369.

Storksdieck, M., & Falk, J. H. (2003). *After 18 months: What determines self-perceived and measured long-term impact of a visit to a science exhibition?* Paper presented at the Visitor Studies Conference, Columbus, OH, July 15–19, 2003.

Storksdieck, M., Falk, J. H., & Witgert, N. (2006, April 3–6). *Why they came and how they benefited: Results of 52 in-depth interviews conducted 18 months after a science center visit.* Paper presented at the NARST 2006 annual meeting, San Francisco, CA.

Strategic Marketing Research. (2000). *Conner Prairie.* Tourism Development Division, Indiana Department of Commerce.

Sykes, M. (1992). "Where learning is child's play": Exhibit evaluation in a children's museum. *Current Trends in Audience Research, 6*, 36–40.

Sykes, M. (1993). Evaluating exhibits for children: What is a meaningful play experience? In D. Thompson, A. Benefield, S. Bitgood, H. Shettel, & R. Williams (Eds.), *Visitor studies: Theory, research and practice: Collected papers from the 1992 Visitor Studies Conference* (Vol. 5, pp. 227–233). Jacksonville, AL: Visitor Studies Association.

Tal, R., Bamberger, Y., & Morag, O. (2005). Guided school visits to natural history museums in Israel: Teachers' roles. *Science Education, 89,* 920–935.

Taylor, P. G. (1996). Reflections on students' conceptions of learning and perceptions of learning environments. *Higher Education Research and Development, 15*(2), 223–237.

Tickle, P. (2005). *Prayer is a place: America's religious landscape observed.* New York: Doubleday.

Tisdal, C. (2004). *Phase 2 summative evaluation of Active Prolonged Engagement at the Exploratorium.* Chicago: Selinda Research.

Tisdal, C., & Perry, D. (2004). *Going APE! at the Exploratorium: Interim summative evaluation report.* Retrieved from www.exploratorium.edu/partner/evaluation .html.

Tokar, S. M. (2003). *Universal design: An optimal approach to the development of hands-on science exhibits in museums.* Unpublished master's thesis, Excelsior College, Albany, NY.

Tomkins, C. (1970). *Merchants and masterpieces. The story of the Metropolitan Museum of Art.* London: Longman.

Trace Center, The. (2004). *Universal design research project.* Retrieved November 2004, from http://trace.wisc.edu/docs/univ design res proj/udrp.htm

Tran, L. U. (2003). *Examining science teacher-student verbal interactions in informal settings.* Williamsburg, VA: International Organisation of Science and Technology.

U.S. Commission on Ocean Policy. (2004). *An ocean blueprint for the 21st century.* Washington, DC: Author.

U.S. Ocean Action Plan. (2004). *The Bush Administration's response to the U.S. Commission on Ocean Policy.* Washington, DC: Author.

Vergeront, J. (2004, March–April). Designed for, not by: The visitor-centered environment. *ASTC Dimensions.*

vom Lehn, D., Heath, C., & Hindmarsh, J. (2001). Exhibiting interaction: Conduct and collaboration in museums and galleries. *Symbolic Interaction, 24*(2), 189–216.

vom Lehn, D., Heath, C., & Hindmarsh, J. (2002). Video-based field studies in museums and galleries. *Visitor Studies Today, 5*(3), 15–23.

Vygotsky, L. (1978). *Mind in society: The development of higher psychological processes.* Cambridge, MA: Harvard University Press.

Vygotsky, L. (1986). *Thought and language.* Cambridge, MA: MIT Press.

Wallace Foundation, The. (n.d.). *"Kickin' it with the Old Masters" at the Baltimore Museum of Art.* Retrieved from www.wallacefoundation.org/WF/Knowledge Center/BaltimoreMuseumDesc.htm.

Walter, C. (2004). Whodunit? The science of solving crime. In K. McLean & C. McEver (Eds.), *Are we there yet? Conversations about best practices in science exhibition development* (pp. 66–69). San Francisco: Exploratorium.

Watson, K., Aubusson, P., Steel, F., & Griffin, J. (2002). A culture of learning in an informal museum setting. *Journal for Australian Research in Early Childhood Education, 9*(1), 125–137.

Watzlawick, P., Weakland, J., & Fisch, R. (1974). *Change: Principles of problem formation and problem resolution*. New York: Norton.

Webb, E. J., Campbell, D. T., Schwartz, R. D., & Sechrest, L. (1966). *Unobtrusive measures: Nonreactive research in the social sciences.* Chicago: Rand McNally.

Weil, S. E. (2002). *Making museums matter.* Washington, DC: Smithsonian Institution Press.

Weil, S. F. (1999). From being about something to being for somebody: The ongoing transformation of the American museum. *Daedalus, 128*(3), 239–258.

Weil, S. F. (2005). A success/failure matrix for museums. *Museum News*, January/February, 36–40.

Wellington, J. J. (1998). Interactive science centers and science education. *Croner's Heads of Science Bulletin* (Vol. 16). Surrey: Croner.

Wellington, J. J., & Osborne, J. (2001). *Language and literacy in science education.* Philadelphia: Open University Press.

Wenger, E. (1998). *Communities of practice: Learning, meaning and identity.* New York: Cambridge University Press.

Wertsch, J. V. (1985). *Vygotsky and the social formation of mind.* Cambridge, MA: Harvard University Press.

Whitney, K., & Associates (2003). *Seeing collection pre-post study final results.* Retrieved from www.exploratorium.edu/partner/evaluation.html.

Williams, D. D. (n.d.). *Educators as inquirers: Using qualitative inquiry.* Retrieved June 2, 2006, from Brigham Young University, Department of Instructional Psychology and Technology, Qualitative Inquiry in Education website, http://msed.byu.edu/ipt/williams/674r/TOC.html.

Wolf, R. L., & Tymitz, B. L. (1979). *A preliminary guide for conducting naturalistic evaluation in studying museum environments.* Unpublished manuscript. Washington, DC: Office of Museum Programs, Smithsonian Institution.

Wolins, I. S., Jensen, N., & Ulzheimer, R. (1992). Children's memories of museum field trips: A qualitative study. *Journal of Museum Education, 17*(2), 17–27.

Women of the West Museum. (2001). *Walk a mile in her shoes: A field guide for creating women's history trails*. Denver, CO: Women of the West Museum.

Wood, D., Bruner, J. S., & Ross, G. (1976). The role of tutoring in problem solving. *Journal of Child Psychology and Psychiatry*, *17*(2), 89–100.

Xanthoudaki, M. (1998). Is it always worth the trip? The contribution of museum and gallery educational programmes to classroom art education. *Cambridge Journal of Education*, *28*(2), 181–195.

Zerafa, G. (2000). *Teacher audience investigation.* Canberra: National Museum of Australia.

Zinchenko, V. P., & Kozulin, A. (1986). The concept of activity in Soviet psychology. *American Psychologist*, *41*, 264–274.

Index

About the Contributors

Sue Allen is director of visitor research and evaluation at the Exploratorium. Over the last decade, she has conducted studies of exhibits and exhibitions, public programs, and teacher professional development courses. Her research focuses on questions that are of interest both to museum practitioners and learning theorists in areas such as scientific inquiry, narratives in science museums, and tools for assessing learning in informal environments. She received her PhD in science education from the SESAME program at UC Berkeley, studying the use of model-based reasoning in geometrical optics.

David Anderson is an associate professor within the department of curriculum studies, University of British Columbia (UBC), Canada. His research has focused broadly on the fields of science education and museum-based learning. His research studies and interests have focused on young childrens' behavior and learning in museum settings, visitors' long-term memories, teacher pre-service training in museum settings, metacognition, and the influence of cultural identity on ways of knowing. Dr. Anderson's consultative and research activities focus on helping museum-based institutions optimize the experiences they provide their visitors and refining the theoretical frameworks through which educational researchers conceptualize and investigate visitor behavior and learning in informal settings.

Tamsin Astor-Jack completed a PhD in cognitive psychology at the Institute of Cognitive Neuroscience, University College London in 2001. Since then, she has become active in science education. She has worked at the Center for Inquiry in Science Teaching and Learning, department of education, Washington University in St. Louis, doing postdoctoral research on professional development in informal science institutions and K–12 science process skills. She has also been a volunteer at the St. Louis Science Center and completed an informal learning certificate with the Center for Informal Learning and Schools. She is currently working as a mother and training to write freelance science education articles for the popular press.

Lynn Baum, program manger of Youth Programs, has been working at the Museum of Science in Boston for over twenty-five years. Although based in the programs division, Ms. Baum has been actively involved in exhibit development and evaluation projects as well. For the last fourteen years Ms. Baum has been overseeing academic and work-based programs for high school students. Throughout the many different projects and programs, the main focus of her work continues to be exploring the impact of teaching and learning in the museum environment.

Jane Blankman-Hetrick is the audience researcher and guest advocate for Conner Prairie Museum in Fishers, Indiana. Ms. Blankman-Hetrick has been with Conner Prairie for six years, conducting both evaluation and research studies to ensure quality experiences for visitors of all ages and backgrounds. She is the co-investigator in Conner Prairie's ongoing research on family learning in the living history setting and has coauthored several articles about living history. Conner Prairie is Ms. Blankman-Hetrick's second career, having spent seventeen years in the telecommunications industry. She has a BA in anthropology and psychology, and is currently completing her master's in sociology.

David Chesebrough is currently the president and CEO of COSI in Columbus, Ohio. CEO and science center innovator over the last twelve years, Dr. Chesebrough was part of the management team that designed and opened the Carnegie Science Center. He is a former educator and served at all levels in schools and colleges for fifteen years prior to joining the science center field. Dr. Chesebrough completed his doctoral research at Duquesne University at Pittsburgh, which focused on museum-community partnerships, and was cosponsored by the American Association of Museums. He also graduated from the University of Pittsburgh completing his bachelor's and master's degrees in physics, math, and science education.

Lynn D. Dierking is Sea Grant Professor of Free-Choice Learning, Oregon State University and vice president for special initiatives, Institute for Learning Innovation, a not-for-profit learning research and development organization focused on understanding, facilitating, and advocating for the role of free-choice learning in a modern knowledge society. Dr. Dierking is internationally recognized for her research on the behavior and learning of children, families, and adults in free-choice learning settings and the development and evaluation of community-based efforts and has published extensively in these areas. She received her PhD in science education at the University of Florida, Gainesville, and has worked in a variety of science learning settings, including museums, schools, community-based organizations, and universities. She serves on the editorial boards of *Science Education* and the *Journal of Museum Management and Curatorship* and has been an advisor for several National Research Council efforts.

Kirsten M. Ellenbogen is the director of evaluation and research in learning at the Science Museum of Minnesota. She began her work in museums as a demonstrator at the Detroit Science Center in 1987, working later in educational programming and exhibit development. Her research focuses on the role of museums in family life, designing exhibits to encourage science conversations, and the ways in which scientific visualization technology can be used to engage the public in exploring scientific data and understanding complex phenomenon. She received her PhD in science education from Vanderbilt University and her BA from the University of Chicago.

John H. Falk is president and founder of the Institute for Learning Innovation, an Annapolis, Maryland-based nonprofit learning research organization; he is also Sea Grant Professor in Free-Choice Learning, Oregon

State University. He is known internationally for his research on free-choice learning. He has authored over one hundred scholarly articles and chapters as well as written and/or edited more than a dozen books, including *Learning from Museums* (2000), *Free-Choice Science Education* (2001), *Lessons without Limit: How Free-Choice Learning Is Transforming Education* (2002), and *Thriving in the Knowledge Age* (2006). Falk earned doctorates in ecology and science education from the University of California, Berkeley.

Susan Foutz is a research associate at the Institute for Learning Innovation. She has worked with science centers, parks, art museums, and other free-choice learning organizations since joining the Institute in 2003. Some of her recent projects have been technology-focused; these include website evaluations, front-end evaluations on cell phone usage in exhibits, and a public television station–school partnership. As the learning innovation initiative coordinator, Susan organized the *In Principle, In Practice* and *Free-Choice Learning and the Environment* initiatives. Susan has a background in sociology and anthropology and received her master's in museum studies from the University of Nebraska, Lincoln.

Cecilia Garibay is an audience research and evaluation consultant, specializing in museums and other free-choice learning environments. As a bicultural/bilingual researcher, Ms. Garibay frames her work in culturally responsive and contextually relevant research and evaluation approaches. Ms. Garibay's research focuses on educational exhibits and programs, particularly projects aimed at reaching underrepresented audiences. As principal of Garibay Group, she regularly consults with institutions on issues of community inclusion and strategies to make exhibitions and programming accessible to multiple and diverse audiences. She is currently conducting research on audience diversity and organizational change in museums.

Des Griffin served as director of the Australian Museum, Sydney, from 1976 until 1998. He has spoken at numerous international meetings and published more than thirty papers on management, leadership, and policy in museums, their role in nature conservation, and on the return of cultural property. He was the first president of Museums Australia (the single association representing museums) from 1993 through 1996 and was a leading player in the establishment of national policies concerning museums in Australia and indigenous peoples. He is presently Gerard Krefft Memorial Fellow at the Australian Museum, Sydney.

Janette Griffin, having taught students from K to 12 in schools, museums, and environmental and science center venues, is now senior lecturer at the University of Technology, Sydney, teaching in undergraduate and postgraduate science education and learning beyond the classroom. Her research and publications investigate ideal conditions for integrated school/museum learning and the complementary roles of teachers and museum educators as well as looking at organizational, social, and educational impacts of museums for visitors and local communities. Janette has worked internationally with leaders in the free-choice learning field and has been engaged for a number of museum learning consultancies.

Joshua Gutwill is senior researcher in the department of visitor research and evaluation at the Exploratorium. His work includes exhibit and program evaluations to improve visitors' experiences and research on learning in informal environments. His focus is on creating opportunities for visitors to engage in self-directed inquiry in science museum settings. Before joining the Exploratorium in 1998, he was the director of assessment and evaluation for a consortium of university faculty creating a new college chemistry curriculum. His driving interest is to use research and evaluation methods to help educators (teachers, exhibit developers, curriculum designers) improve their practice.

Jeff Hayward founded People, Places & Design Research in 1984 to help museums understand their audiences. By professional training and practice, he is an environmental psychologist specializing in audience research (visitor characteristics, interests, patterns of behavior and experience, knowledge and understanding of interpretive concepts, etc.), and has conducted about 500 studies for over 150 cultural-interpretive nonprofits (museums, aquariums, botanic gardens, historic sites, etc.). Hayward considers audience research to be much more than basic documentation; he sees it as a management and planning tool that can inform creative and strategic decision-making for any size institution.

Joe E. Heimlich is a professor of environmental education and interpretation at Ohio State University and a senior research associate with the Institute for Learning Innovation. He has been engaged in the arena of environmental education free-choice learning for fourteen years as a professor and before that as an extension associate with OSU Extension. Dr. Heimlich holds a doctorate in adult education and learning theory from Ohio State University; a BA in communication arts, theatre, and dance, and a MA in policy education. Joe is a past president of NAAEE, and is active nationally and internationally as an evaluator of environmental education programs.

Gretchen Jennings is well known as a museum educator and exhibition developer and coordinator. She has served most recently as director of education for interpretation and visitor experience at the Smithsonian's National Museum of American History. She has been a project director or senior staff member on several major traveling exhibitions, including *Invention at Play, Secrets of Aging*, and *Psychology: Understanding Ourselves, Understanding Each Other*. Both *Invention at Play* and *Psychology* received awards of exhibition excellence from the American Association of Museums. Ms. Jennings has published and spoken widely on topics that include cultural diversity, exhibition development, project management for traveling exhibitions, and the creation of appropriate museum spaces for young children. From 1996 to 1999 she served as editor in chief for the *Journal of Museum Education*. In May 2007 she became managing editor of *The Exhibitionist*, the journal of the National Association of Museum Exhibition (NAME). Currently Ms. Jennings consults with museums on a range of topics. Her email is gretchenjennings@rcn.com.

Julie I. Johnson is the Science Museum of Minnesota's first John Roe Chair of Museum Leadership. She helps to further the implementation of museum goals, and provides support and leadership in the areas of planning, programming, personnel development, and collaboration. From 2003–2005 Johnson was program officer at the National Science Foundation (NSF) in the elementary, secondary, and informal education division. Concurrent with the assignment at NSF, Johnson was the executive vice president and chief operating officer for the former New Jersey State Aquarium. There she spearheaded the development of programs involving local youth and was project director and principal investigator for the PISEC *Family Science Learning Study* and *Families Exploring Science Together*. She serves on the board of the Visitor Studies Association and is on the faculty for the Getty Leadership Institute's *Museums Leaders: the Next Generation.*

David J. Johnston is currently visiting professor at the office for research and development, Curtin University of Technology, Perth, Western Australia, where he is developing a course for doctoral students on leadership and communications for commercializing research. Previously, Dr. Johnston was director, Centre for the Advancement of University Teaching, University of Hong Kong, where he was a leader and consultant for implementation of innovative practice in universities throughout Asia. He was the founding president of the Asia-Pacific Problem-Based Learning Association. His research has mainly been about determining outcomes of visits to free-choice learning environments and the impact of problem-based learning in education for the professions.

Christine Klein, independent consultant, conducts research and evaluation at the interface between informal and formal science education and into the nature of collaboration among informal science institutions, universities, community organizations, and schools. Previously, Dr. Klein coordinated the partnership between the St. Louis Science Center and Compton-Drew, their museum school. Later at the Science Center and then at Washington University in St. Louis, Dr. Klein served as principal investigator for the Center for Inquiry in Science Teaching and Learning, a collaboration among higher education, informal science, and school districts to improve science education through professional development for educators and research in science education.

Emlyn H. Koster, in 1996, after leadership of the Ontario Science Centre and Royal Tyrrell Museum in Canada, was appointed president and CEO of Liberty Science Center. Situated next to the Statue of Liberty and Ellis Island, this institution reopens in 2007 after a $104 million expansion and renewal. Author of more than one hundred articles and an invited speaker globally, he has advocated since the early 1990s for the museum field to embrace external responsibilities. Representing the science center field, his board appointments include the American Association for the Advancement of Science, Institute for Learning Innovation, and the International Council of Museums / American Association of Museums. His honors include the John Cotton Dana Award from the New Jersey Association of Museums and elected presidencies of the Geological Association of Canada and Giant Screen Theater Association.

Jessica J. Luke is senior researcher at the Institute for Learning Innovation, an Annapolis, Maryland-based not-for-profit learning research and development organization. She has ten years of experience studying learning in museums, and has conducted myriad research and evaluation studies in support of program and exhibition development at art museums, children's museums, and science centers across the country. Ms. Luke's research is focused on museums and community, in particular related to issues of youth development and parent involvement. She has published various articles and book chapters on these topics. In addition, Ms. Luke is an adjunct faculty member at George Washington University, where she co-teaches a graduate course on museum evaluation in the department of museum studies.

Laura M. W. Martin, director of science interpretation at the Arizona Science Center, was previously executive vice president for experiences at the Phoenix Zoo. She has been a researcher, teacher, and program administrator in formal and informal settings. In the museum setting, she has served as vice president of education and research at the Arizona Science Center and research director at the Center of Teaching and Learning at the Exploratorium. She teaches in the museum studies program of Arizona State University and is a member of the National Academies Committee on Learning Science in Informal Environments.

Mary Ellen Munley has twenty-five years experience as a museum educator and administrator. Recently, as director of education at The Field Museum, Chicago, she oversaw a redesign of all education programs and helped form Museums and Public Schools (MAPS), a partnership with Chicago Public Schools. Ms. Munley is principal of MEM and Associates, a consulting firm in Chicago dedicated to enhancing the role of museums in the lives of people and communities. She is a recipient of the American

Association of Museums award for excellence in the practice of museum education, and president of the Visitor Studies Association.

Jeffrey Patchen is president and CEO of The Children's Museum of Indianapolis, and has directed the museum's vision and long-range strategic efforts since 1999. From 1996 to 1999, Dr. Patchen served as senior program officer for national programs for The Getty Education Institute for the Arts, an operating program of the J. Paul Getty Trust in Los Angeles, California. From 1990–1996, Patchen was the Lyndhurst Endowed Chair of Excellence in Arts Education at the University of Tennessee at Chattanooga where he directed the Southeast Center for Education in the Arts. From 1984–1990, Dr. Patchen served as the state arts consultant for the Indiana Department of Education.

Erminia Pedretti is associate professor of science education at the Ontario Institute for Studies in Education of the University of Toronto. She is also director of the Centre for Studies in Science, Mathematics and Technology Education. Her research interests include: science, technology, society and environmental education, action research, teacher professional development, and learning science in non-school settings. Recent publications include studies of issues-based exhibitions and learning in science centers, the development and implementation of multimedia case studies in teacher education programs, and a co-edited book (with S. Alsop and L. Bencze) titled *Analyzing Exemplary Science Teaching*.

Deborah L. Perry, PhD has been director of Selinda Research Associates, Inc. since 1989. With a background in instructional systems technology, she has conducted research on the role of intrinsic motivation in informal educational settings focusing on the question "What makes learning fun?" with a particular emphasis on the social dimensions of the visitor experience. She has extensive experience in free-choice learning, and has consulted with museums and other organizations throughout the United States, Canada, and in the United Kingdom. Her work includes evaluating exhibitions and programs, planning and presenting workshops, and incorporating instructional development principles into the design of visitor experiences. She specializes in naturalistic inquiry and uses a variety of qualitative and quantitative methods.

Anne Grimes Rand is deputy director of the USS Constitution Museum, a free museum located on Boston's historic Freedom Trail; she is responsible for the visitor experience for the 250,000 people who visit the museum each year. Rand enjoys creating interactive displays that encourage family learning by engaging visitors of all ages in hands-on and minds-on history. She oversees the museum's exhibits, programs, and educational efforts both at the museum and on the road. Rand served as curator of the USS Constitution Museum for twelve years, and she is an avid sailor who prefers to be afloat.

Christine A. Reich is manager of informal education research and evaluation at the Museum of Science, Boston. She began her career as an exhibit planner, working to engage visitors with a broad range of abilities and disabilities in science inquiry. As a researcher, her efforts continue to focus on universal design. In addition to her work at the museum, Ms. Reich instructed visitor studies at Harvard University, serves on the editorial board of ASTC's *Dimensions*, and is the secretary/treasurer of CARE. She has BS in engineering from Cornell University, a certificate in museum studies from Harvard University, a master's of education from Lesley University, and is working toward a PhD in education at Boston College.

Léonie J. Rennie is professor of science and technology education at Curtin University of Technology, Perth, Western Australia. Her background is in science teaching and curriculum, and her scholarly publications include over 150 books, chapters, and refereed journal articles. Dr. Rennie has conducted research about learning and the communication of science in school and out-of-school settings. She coedited a special edition of the *Journal of Research in Science Teaching* devoted to learning science in informal contexts, has papers in special editions of the *International Journal of Science Education* and *Science Education*, and is author of the chapter on out-of-school learning for the forthcoming *Handbook of Research in Science Education*.

Randy C. Roberts has more than twenty years of experience in museum education, administration, and strategy development. She is the manager of the Visitor Studies Association, an international professional organization focusing on all facets of the visitor experience in free-choice learning settings. She also serves as a senior associate with MEM and Associates. Randy holds a master's in public administration from Ohio University.

Shawn Rowe joined the staff of Sea Grant Extension and the department of science and math education at Oregon State University after receiving his PhD from Washington University in St. Louis. His research combines methods for studying learning outside of school with tools for analyzing how groups make sense out of what they say and do. Along with Dr. Olga Rowe, he has carried out research in history and science museums both in the United States and in Ukraine. His current position combines academic work on free-choice learning with practice-based research in the OSU Hatfield Marine Science Visitor Center in Newport, OR, and teaching and advising.

Jerry Schubel joined the Aquarium of the Pacific as president and CEO in June 2002. His emphasis is to distinguish the Aquarium of the Pacific as a free-choice learning institution in addition to its status as a world-class aquarium. He has served on numerous National Research Council commissions, committees, and boards and chaired the Marine Board. He is past chair of the National Sea Grant Review Panel and has served on the NSF's Education and Human Resources Advisory Council. He is president emeritus of the New England Aquarium, where he was president and CEO from 1994 to 2001. He was dean and director of Stony Brook University's Marine Sciences Research Center from 1974 to 1994, and provost for three of those years. He has written extensively for scientific journals and for general audiences.

Beverly Sheppard is executive director of the Institute for Learning Innovation. Her work in developing and supporting partnerships is extensive. As acting director of the Institute of Museum and Library Services, she enabled partnerships between museum and library partners and initiated the collaborative project *A Partnership for a Nation of Learners*. Her career includes more than twenty years in museum settings, serving most recently as president and CEO of Old Sturbridge Village in Massachusetts. Sheppard is coauthor, with John H. Falk, of *Thriving in the Knowledge Age* and author/editor of *Building School and Museum Partnerships*.

Barbara Soren is an educator who has been working with museums and science centers, performing arts organizations, community organizations, health care facilities, and schools since the mid-1970s. Her academic background includes a PhD in education from the University of Toronto and a master of science in teaching from McMaster University. She is intrigued with how people grow and learn throughout their lives in rich and meaningful contexts. Her consulting work, research and teaching have focused on: lifelong learning and human development; understanding experiences individuals have in museums and arts and community organizations; and evaluating quality online user experiences in the cultural and health sectors.

Michael Spock is a research fellow at the Chapin Hall Center for Children at the University of Chicago where he has been a student of learning in museums and other informal settings; he also holds an appointment as museum-scholar-in-residence at the Chicago History Museum. From 1986 to 1994, Dr. Spock served as vice president for public programs at the Field Museum, and before coming to the Field was director of the Boston Children's Museum for twenty-three years. Spock has been an advisor to museums and funders, and was instrumental in the founding, or served in the leadership, of national and regional museum and umbrella cultural organizations.

Martin Storksdieck is a senior researcher at the Institute for Learning Innovation where his research and evaluation studies focus on the public's understanding of science and research, factors that influence cognitive gains in free-choice learning environments, long-term and alternative outcomes of museum visits, determinants of behavioral and attitudinal change, and school field trips to free-choice learning environments. Prior to joining the institute, Dr. Storksdieck worked as an environmental management consultant, was a science educator and producer with a planetarium in Germany, and served as editor, host, and producer for a weekly environmental news broadcast.

David Ucko heads the informal science education program at the National Science Foundation. His experience includes: president, Museums+*more* LLC; president, Science City at Union Station (Kansas City Museum);

deputy director, California Museum of Science & Industry; vice president, Chicago's Museum of Science & Industry, and associate professor of chemistry, Antioch College. Ucko has served as presidential appointee to the National Museum Services Board, Institute of Museum and Library Services and has been recognized as an AAAS Fellow and Woodrow Wilson Fellow. He is the author of two college chemistry textbooks. Ucko received his PhD from M.I.T. and BA from Columbia.

Robert "Mac" West, president of Informal Learning Experiences, Inc., has been in the museum business for over thirty years. After earning his PhD in vertebrate paleontology from the University of Chicago, he taught biology at Adelphi University, Garden City, New York, before moving to become curator of geology at the Milwaukee Public Museum. That was followed by appointments as director of The Carnegie Museum of Natural History in Pittsburgh, and Cranbrook Institute of Science in Bloomfield Hills, Michigan. He founded Informal Learning Experiences (ILE) in 1992. ILE consults broadly in the areas of institutional planning, development, and sustainability. It publishes *The Informal Learning Review* and maintains its online searchable Traveling Exhibitions Database.

Kimberlee L. Kiehl Whaley is currently vice president for Learning & Research Partnerships at COSI. In this role she has oversight on student and teacher programming, outreach programming, early childhood programming, and research. She was previously associate professor of human development and family science and state extension specialist at Ohio State University, Columbus, and curriculum coordinator for the A. Sophie Rogers Laboratory for Child and Family Studies. She also served as associate dean for academic programs, College of Human Ecology, and has extensive publications in professional journals. Dr. Whaley originally came to COSI on "special appointment" as associate vice president for early childhood education for COSI Columbus on loan from Ohio State. Dr. Whaley received a BS in speech pathology and audiology in 1981 at SUNY Geneseo, graduating magna cum laude. The following year (1982) she completed her master's degree at The College of St. Rose, department of special education. Dr. Whaley received her PhD in curriculum and instruction from Penn State University in 1990. She continues to teach at Ohio State University in the department of education and human ecology.